OECD Compendium of Agri-environmental Indicators

Please cite this publication as:
OECD (2013), *OECD Compendium of Agri-environmental Indicators*, OECD Publishing.
http://dx.doi.org/10.1787/9789264181151-en

ISBN 978-92-64-18115-1 (print)
ISBN 978-92-64-18621-7 (PDF)

Photo credits: Cover © Shutterstock/Eky Studio

Corrigenda to OECD publications may be found on line at: *www.oecd.org/publishing/corrigenda*.

Foreword

The OECD Compendium of Agri-environmental Indicators *provides the latest and most comprehensive set of agri-environmental indicators (AEIs) across 34 OECD countries from 1990 to 2010. It builds on 20 years of OECD work on developing AEIs (see* www.oecd.org/tad/sustainable-agriculture/agri-environmentalindicators.htm*). The AEIs in the report seek to: describe the current state and trends of environmental conditions in agriculture; highlight where "hot spots" are emerging; compare trends in performance across time and between countries; and provide a set of indicators and database which can be drawn on for: policy monitoring and evaluation; projecting future trends; and for developing green growth indicators.*

The report recognises the formidable problems involved in developing a comparative set of agri-environmental data. In most countries gathering such data only began in the early 1990s. Methodologies to measure the environmental performance of agriculture are not well-established in all cases. National average data often conceal significant ranges reflecting local site-specific values. A vast amount of data is potentially of interest but the attempt here, even though partial, is to focus on those that are of use to policy makers.

But even given these caveats this report provides a wealth of data and information for policy makers, researchers and stakeholders wanting to know, explore and analyse agriculture's impact on the environment, whether through modelling efforts, or through simply looking at time or cross-country series data. Moreover, the project has striven to develop broadly agreed methodologies of measurement that can be used at the national, local or farm levels.

The project was carried out under the auspices of the OECD Joint Working Party on Agriculture and the Environment (JWPAE), of the Committee for Agriculture and the Environment Policy Committee. The JWPAE approved the report for publication in January 2013.

The OECD wishes to acknowledge the contribution of member countries in the preparation of this report, especially through expert comment on the text and provision of data in the study. OECD would also like to thank Eurostat who helped in variously providing information, especially Johan Selenius and Annemiek Kremer, as well as the European Bird Census Council, and the Secretariats of the UN Economic Commission for Europe, UN Environment Programme and the UN Framework Convention on Climate Change.

The principal authors of this report were Julien Hardelin and Kevin Parris, both economists in the Environment Division of the Trade and Agriculture Directorate headed by Dale Andrew (Kevin Parris has now retired from the OECD). Within the Secretariat, many colleagues from the Trade and Agriculture Directorate and the Environment Directorate contributed to the report, in particular, Françoise Bénicourt, Eric Espinasse, Frano Ilicic, Jussi Lankoski, Theresa Poincet, Véronique de Saint-Martin, Noura Takrouri-Jolly and Tetsuya Uetake. Valuable assistance was also provided by the OECD Translation Service and the Public Affairs and Communications Directorate, plus a number of former OECD colleagues, including Wilfrid Legg and Evelyne Misak.

Table of contents

Tables

Figures

Executive summary

The recent environmental performance of agriculture provides some encouraging signs that agriculture is capable of meeting future environmental challenges (Figure 0.1). Evidence for OECD countries from 1990 to 2010 show improvements have been made in nutrient, pesticide, energy and water management, using less of these inputs per unit volume of output. Enhanced environmental performance has also flowed from the more widespread adoption of environmentally beneficial practices by farmers, such as conservation tillage, improved manure storage, soil nutrient testing, and drip irrigation.

Figure 0.1. **Key agri-environmental indicators, OECD average, 1990-2010**

% annual growth rates 1990-92 to 1998-2000 and 1998-2000 to 2008-10

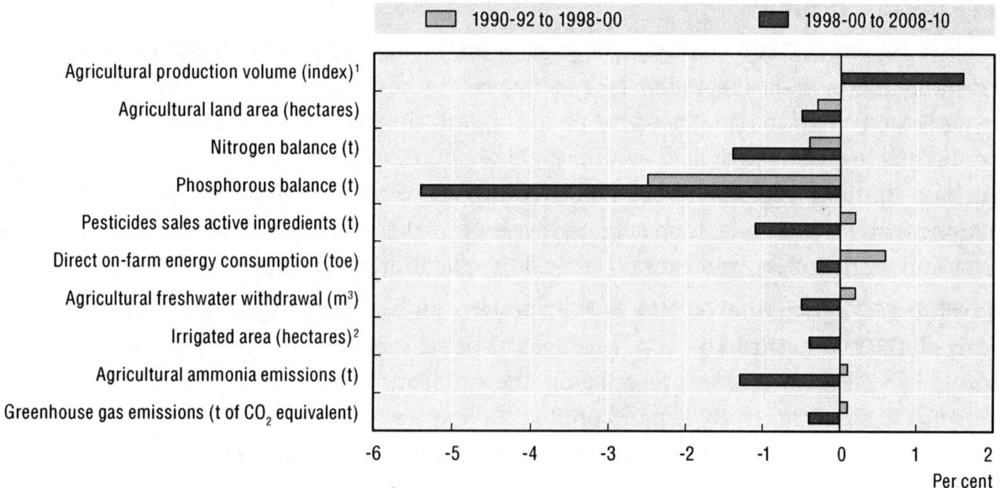

Notes: t: tonnes; toe: tonnes oil equivalent; m^3: cubic meters; CO_2: carbon dioxide.

The OECD total average for the indicators listed here is the average for 34 member countries, except (figure in brackets show the number of OECD countries included in the average calculation): nitrogen and phosphorus balance (31); pesticide sales (29); on-farm energy consumption (32); freshwater withdrawals (24); irrigated area (21); and ammonia emissions (26).

1. For technical reasons, the OECD agricultural production volume annual average growth rate is not calculated for the period 1990-92 to 1998-2000.
2. The annual growth rate for irrigated area between 1990-92 to 1998-2000 was less than 0.1% per annum.

StatLink http://dx.doi.org/10.1787/888932792331

Agriculture has a significant position with respect to the environment (Table 1.1), especially due to the amount of land and water it uses, in contrast to a much smaller role in the overall economy (e.g. share in employment and GDP). Agriculture is recognised to produce both positive (e.g. carbon sequestration) and negative environmental externalities (e.g. water pollution), that are not reflected in agricultural GDP as there are few markets for these externalities. The value of the positive and negative externalities generated by agriculture are likely to be substantial, but no comprehensive monetary assessment of these costs and benefits currently exists.

The positive signs of environmental improvements partly originate from the better integration of environmental issues in farmers' decision making since the early 1990s. This reflects a combination of more stringent environmental regulations, increases in agri-environmental payments, and development of measures, such as market-based instruments, collective action, and technical assistance and research. Also over the past decade the slowdown in the growth of agricultural production compared to the 1990s, for most OECD countries, has in most cases tended to lower the growth in farm input use (nutrients, pesticides, energy and water) and emissions from livestock (ammonia, methane), which has enhanced environmental quality.

The total payments from OECD taxpayers to agricultural producers to generate environmental benefits and reduce environmental costs, have risen substantially since the early 1990s and now run into billions of dollars annually, although no exact estimate is available. Identifying the extent to which these budgetary payments over the past 20 years have shaped the environmental performance of OECD agriculture is complex and not fully understood. This is because payments are only one of the drivers affecting environmental change, as developments in other policies, the economy, markets, technology, knowledge, societal expectations, and the natural environment all play a part in shaping environmental outcomes in agriculture.

For some regions in OECD countries progress in improving the environmental performance of agriculture has been disappointing. More effort is required from farmers, policy makers and the agro-food chain to address water pollution and the decline in farmland breeding bird populations in these regions. Absolute levels of pollution also remain a challenge, as they continue in many regions across OECD countries to exert significant pressure on the environment. For example, high surplus levels of nitrogen and phosphorus leading to soil, water and air pollution, and excess pesticide application causing groundwater pollution.

The OECD-FAO agricultural outlook to 2021 projects an increase in agricultural production for nearly all OECD countries. Under a "business as usual scenario" the projected expansion in production could increase pressure on the environment. This poses a major policy challenge to simultaneously expand production to meet global food security demands and at the same time reduce the environmental costs and encourage the environmental benefits associated with agriculture.

To address the twin policy challenges of ensuring global food security and improving environmental performance will require raising the environmental and resource productivity of agriculture; enhancing land management practices; minimising pollution discharges; curtailing damage to biodiversity; and strengthening policies that avoid the use of production and input subsidies damaging to the environment.

PART I

Background and scope of the report

Chapter 1

Overview

This chapter presents the objectives, scope, and structure of the report. It examines the coverage and methodology relating to the data underlying the OECD agri-environmental indicators, their main caveats and limitations, and an assessment according to the four OECD criteria of: policy relevance, analytical soundness, measurability and interpretability. The chapter also provides a summary of the main trends of the environmental performances of agriculture in the OECD area since 1990.

The **OECD *Compendium of Agri-environmental Indicators*** provides the latest and most comprehensive set of agri-environmental indicators (AEIs) across 34 OECD countries from 1990 to 2010. It builds on 20 years of OECD work on developing AEIs (see *www.oecd.org/tad/sustainable-agriculture/agri-environmentalindicators.htm*). The AEIs in this Report seek to: describe the current state and trends of environmental conditions in agriculture; highlight where "hot spots" are emerging; compare trends in performance across time and between countries; and provide a set of indicators and database which can be drawn on for policy monitoring, evaluation, in projecting future trends, and for developing green growth indicators.

Agri-environmental monitoring has improved for most countries, but there remain many gaps in the coverage and quality of the indicators in this report. Information on pesticides is illustrative: most countries only record sales data and not actual on-farm use; the distinction between agricultural and other uses (e.g. home gardens) is rarely monitored; while indicators revealing the consequences of pesticide use in agriculture for human health, water quality and wildlife are lacking in most OECD countries. Without adequate information to monitor the environmental perfomance of agriculture, it is difficult to assess the efficiency and effectiveness of policies that seek to deliver environmental benefits and reduce costs.

In reading this report, it is important for the reader to take into account a number of caveats and limitations when making assessment of AEIs over time and across countries, as summarised below.

- *Definitions and methodologies for calculating indicators* are standardised in most cases but not all, in particular indicators for soil erosion and water quality are not fully standardised.
- *Scientific understanding of biophysical relationships and their interaction with farming activities are poor in a number of areas*, such as the pathways and extent of pollutants into groundwater.
- *Data availability, quality and comparability* are as far as possible complete, consistent and harmonised across the various indicators, but deficiencies and data gaps remain.
- *Spatial aggregation* of indicators is given at the national level, but for many indicators (e.g. nutrient balances, soil erosion, water quality) this can mask significant variations at the regional level.
- *Trends in indicators*, rather than absolute levels, are important for comparative purposes across countries for many indicator areas, especially as local site specific conditions can vary considerably.
- *Agriculture's contribution to specific environmental impacts* can be difficult to isolate, especially where the impact of other activities are important (e.g. contribution of industry to water pollution).
- *Environmental improvement or deterioration* is in most cases clearly revealed by the direction of change in the indicators, but in some cases changes can be ambiguous, such as changes in land cover.

- *Baselines, threshold levels or targets for indicators* are generally not used to assess indicator trends in the report as these may vary between countries due to different national circumstances.

These caveats and limitations, however, need to be viewed in a broader context. In many cases they also apply to other socio-economic indicators regularly used by policy makers. For example, there can be wide regional variations around national averages of socio-economic indicators (e.g. unemployment, average family income), and methodological and data deficiency problems are also common (e.g. wealth distribution). Work on establishing AEIs is also relatively recent compared with the longer history of developing economic indicators, such as GDP. Measuring causal linkages between the biophysical environment and human activities is also difficult given many environmental effects do not have market (monetary) valuations, and are often measured using different physical units (e.g. tonnes, CO_2 equivalents, ozone depleting potential).

1.1. Agri-environmental performance in specific areas

Production and land

The relationship between changes in the volume of agricultural production and agricultural land area can provide a broad indication of the environmental performance of agriculture. Increases in agricultural production and land use often signify greater pressure on the environment, as may the intensification of production on a reduced area farmed. Environmental pressure, however, will depend on the extent to which farming practices limit the pressures, such as improving resource use efficiency.

Growth in the volume of OECD agricultural production slowed over the decade from 2000 to 2010 compared to the 1990s (Figure 0.1). With the ***constant decline in the total OECD agricultural land area*** since the 1990s (Figure 0.1), agriculture has intensified production on a reduced land area by raising crop yields and animal stocking densities. Production has grown most rapidly for the major OECD agricultural exporting countries, such as **Australia**, **Canada**, **Chile**, **Israel**, **New Zealand**, **Turkey** and **United States**, compared to **European OECD countries** (except notably **Spain**) and **Japan**. Where production has grown rapidly this has frequently heightened environmental pressure, mainly linked to greater use of inputs (e.g. fertilisers, pesticides and water) and intensification of livestock operations.

The OECD area of agricultural land declined continuously over the period 1990 to 2010 (Figure 0.1). But despite the overall trend to remove land from agricultural use it remains the major land use for many countries, with agriculture representing over 40% of the total land area for two-thirds of OECD countries by 2008-10 (Table 1.1). Agricultural land area has expanded in only six OECD countries between 2000 and 2010. Where this has occurred it has mainly involved countries with rapidly expanding agricultural sectors (e.g. **Canada**, **Chile**) or cases where better reporting of land farmed is partly due to improved registration systems linked to requirements for payments under some agri-environmental schemes and more generally agricultural support policies, especially in EU countries (e.g. **Finland**, **Luxembourg**, **Sweden**).

The share of agricultural land under certified organic farming remains very low across OECD countries, below 2% for the OECD average 2008-10 (Table 1.1). But this masks substantial variation across countries with shares tending to be higher than the OECD average in mainly **EU** countries, and below the average for most non-EU countries. To some extent this reflects varying policy environments, for example, with organic conversion

Table 1.1. **Primary agriculture in the economy and the environment, OECD countries, 2008-10**

Percentage of OECD primary agriculture in total	OECD average	Range of values (minimum to maximum)
• GDP	2.6%	0.3 to 9.2%
• Land area	36%	3 to 72%
• Certified organic farm area as a share of total agricultural area	1.9%	0.01 to 15.6%
• Nutrient balances (surpluses and deficits)		
Nitrogen, kg per hectare of agricultural land	63 kg/ha	1 to 228 kg/ha
Phosphorus, kg per hectare of agricultural land	6 kg/ha	-10 to 49 kg/ha
• Pesticide sales	70%	65 to 80%
• Energy consumption	1.6%	0.4 to 6.3%
• Water withdrawals	44%	0.2 to 89%
• Irrigated land area share in total agricultural area	4%	0.4 to 54%
❖ Water pollutants, *of which:*		
Nitrates in surface water	..	33 to 82%
Nitrates in groundwater[1]	..	1 to 34%
Nitrates in coastal water	..	35 to 78%
Phosphorus in surface water	..	17 to 70%
Phosphorus in coastal water	..	23 to 50%
Pesticides in surface water[1]	..	0 to 75%
Pesticides in groundwater[1]	..	0 to 25%
• Ammonia emissions	91%	82 to 98%
• Greenhouse gas emissions	8%	2 to 46%
Of which: Nitrous oxide emissions	75%	..
Methane emissions	38%	..
• Share of OECD methyl bromide use in world total:		
Ozone depleting products	5%	..
Methyl bromide use	46%	..

..: Not available.

Note: The data in this table should be interpreted as approximate values rather than precise values, and for some indicators includes forestry and fisheries. For full notes and sources, consult the website below.

1. Share of monitoring sites exceeding recommended drinking water threshold limits.

Source: *OECD Agri-environmental Indicator Database, www.oecd.org/tad/sustainable-agriculture/agri-environmentalindicators.htm.*

StatLink http://dx.doi.org/10.1787/888932793395

payments provided to EU farmers, but not available to farmers in countries such as **Australia**, **Canada**, **Chile**, **Israel** and **New Zealand**. This is also reflected in the variable growth in organic farming from 2002 to 2010, with growth more rapid in mainly European OECD countries (e.g. **Austria**, **Czech and Slovak Republics**, **Estonia** and **Sweden**), and less rapid in largely non-European OECD countries (e.g. **Japan**, **Mexico**).

Organic farming systems usually involve practices that maintain or improve the physical, chemical and biological conditions of soil, compared to other farming systems. Organic farming practices can also bring other benefits, such as to water quality by not using synthetic pesticides, as well as providing other ecosystem services, for example, carbon sequestration and enhanced biodiversity. There are situations, however, where intensive management within organic farming can lead to livestock manure, for example, being applied in excess of requirements. Organic farming also often involves increased tillage to manage weeds (in the absence of pesticides), which may increase soil erosion.

Some 20% of the total OECD arable and permanent cropland area is sown to transgenic crops in 2008-10 (sometimes referred to as *genetically modified crops*). The **United States** dominates OECD commercial production of transgenic crops. Regulations in **European OECD countries** and **Korea**, prevent the commercial exploitation of these crops, with only small areas sown for experimental purposes. The OECD area of transgenic crops has grown rapidly since the mid-1990s, especially in **Canada** and the **United States**, dominated by herbicide tolerant crops (soybean, maize, canola, and cotton). OECD countries account for slightly more than half of the world global planted area of transgenic crops, but countries such as Argentina, Brazil, China and India have expanded the area of these crops substantially.

The development of transgenic crops has led to ongoing discussions and debate on the potential environmental costs and benefits of using these crops, as well as safety for human health. For example concerns have been raised over the possibility of genetic mingling of traditional species and wild relatives, such as maize in **Mexico** recognised as a "Vavilov" centre, which is an area where crops were first domesticated and have evolved over several thousand years, as is the case for maize. At the same time some researchers view transgenic crops as bringing benefits in terms, for example, of reducing pesticide use or providing crops with water saving traits.

Nutrient balances *(i.e. the balance between nitrogen and phosphorus inputs, largely inorganic fertilisers and livestock manure, and outputs, the uptake of nutrients by crops and pasture).*

Inputs of nutrients, such as nitrogen and phosphorus, are necessary in farming systems as they are critical in maintaining and raising crop and forage productivity. Where nutrients are in deficit soil fertility and output will decline, while with an excess of nutrients necessary for plant growth there is a risk of polluting soil, air, and water (eutrophication). OECD agriculture is a significant source of nitrogen and phosphorus entering the environment as there is in most cases a surplus of nutrients compared to plant requirements.

Overall OECD agricultural nitrogen and phosphorus surpluses have been on a constant downward trend from 1990 to 2009, measured in tonnes of nutrients and in terms of nutrient surpluses per hectare of agricultural land (Figure 0.1). The rate of reduction in OECD nutrient surpluses was more pronounced over the 2000s compared to the 1990s, especially for phosphorus. These trends reflect both overall improvements in nutrient use efficiency by farmers, and the slower growth in agricultural production for many countries over the 2000s, and for phosphorus the growing realisation by farmers that high levels of accumulated phosphorus are stored in their soils.

The lowering of nutrient surpluses has reduced the risk of environmental pressure on soil, water and air. Despite this improvement ***nutrient intensity levels per hectare of agricultural land remain at high levels in terms of their potential to cause environmental damage for most OECD countries***, with OECD averages for 2007-09 of 63 kg nitrogen (N) per hectare and 6 kg phosphorus (P) per hectare. But there are sizable variations within and across countries in terms of the intensity and trends of nutrient surpluses (Table 1.1). Background (or natural) loss of nitrogen is typically estimated at around 1-2 kg N/ha from electrical sources and other sources, while for phosphorus this figure is about 0.1 kg P/ha depending on underlying conditions in sediment and rocks.

Pesticide sales and associated risks to human and environmental health

Pesticides are major inputs for agriculture that facilitate lowering the risks of yield losses. As agriculture is the major user of pesticides (Table 1.1), it also poses a significant source of risk to pollution of water systems and is of concern for human and wildlife health and the functioning of ecosystems. As a result there is an extensive range of policy instruments used by OECD countries to address human and ecosystem health concerns and pollution of water associated with the application and disposal of pesticides in agriculture.

Overall OECD pesticide sales (volume of active ingredients) diminished by over 1% per annum over the period 2000-10, which contrasts to a 0.2% per annum increase over the 1990s (Figure 0.1). For a few countries, however, pesticide sales have been rising over the 2000s, largely driven by increasing crop production, especially horticulture and vines, which in part explains recent increases in pesticide sales for Chile, Estonia, Finland, Hungary, Iceland, Mexico, New Zealand, Poland, Spain and Turkey.

There is evidence that for a growing number of countries ***the volume of crop production has been increasing at a faster rate over the period since 2000 than the change in pesticide sales***. The apparent efficiency improvements in the use of pesticides in crop production can be explained by a number of factors, varying in importance between countries, including: farmer education and training for example, calibrating and ensuring pesticide sprayers are working properly; the overall decoupling of support from production and input related support; the use of payments to encourage adoption of beneficial pest management practices; pesticide taxes; the use of new pesticide products in lower and more targeted doses; and the expansion in organic farming.

There are no comparable cross country data on the risks to human health and the environment from the use of pesticides in agriculture. This is despite considerable research in the area and also that a few countries have developed their own human health and environmental pesticide risk indicators. The lack of data and indicators on pesticide risks is further compounded in terms of poor knowledge and information on the health and environmental effects with the release of mixtures of pesticides rather than a single pesticide product, and the interaction in the environment between pesticides and other chemical contaminants (e.g. veterinary medicines, human pharmaceuticals, personal care products and industrial chemicals). In most OECD countries, however, regulatory processes are removing older, more persistent and toxic pesticides from the market, such as DDT.

On-farm energy consumption and production of biofuels from agricultural feedstocks

Agriculture can play a double role in relation to energy, both as a consumer to power farming operations and also as a producer of bioenergy, including biofuels using agricultural feedstocks. Support to agricultural energy use is widespread across OECD countries and typically involves reducing the standard rate of fuel tax for on-farm consumption. Support is also common for biofuels by providing a combination of mandates, tax incentives and payments for the production and use of biofuels.

Direct on-farm energy consumption declined over the period 2000 to 2010 compared to an increasing trend over the 1990s (Figure 0.1). To a large extent this reflects the slowdown in OECD agricultural production over the same period, although the share of primary agriculture in total national energy consumption is extremely low for most OECD countries (Table 1.1). Improving energy efficiency in primary agriculture is taking on an increasingly important role for nearly all countries, not only in terms of the need to reduce overall farm operating costs, but also as part of national programmes to lower greenhouse gas emissions from the use of fossil fuels.

While production of biofuels has a long history in some countries, for most countries production has expanded rapidly over the period from 2000 to 2010. Bioethanol production dominates biofuel production in OECD countries, accounting for 77% of total OECD biofuel production in 2008-10, converted in energy terms. The feedstocks to produce biofuels in OECD countries are largely maize in the United States to produce bioethanol, and in the European Union rapeseed oil is mainly used to produce biodiesel.

A key conclusion from most studies on the ***consequences of biofuel feedstock production on the environment*** is that in general feedstocks from annual arable crops (e.g. maize, rapeseed), can have a more damaging impact on the environment than second generation feedstocks (e.g. reed canary grass, short rotation woodlands, farm waste). Another important conclusion is that the location of production and the type of tillage practice, crop rotation system and other farm management practices used in producing feedstocks for biofuel production, will also greatly influence environmental outcomes.

Soil erosion from water and wind

Soil erosion, mainly through water and wind processes, is one of the most widespread forms of soil degradation across OECD countries. Agricultural soils provide two key functions: to support production (notably agriculture but also forestry, etc.); and environmental functions, such as water filtration and conservation, carbon sequestration, and as a reservoir for biodiversity. Soil conservation, including lowering soil erosion risks, is crucial to ensure these soil functions can be maintained.

Over 20% of the agricultural land area is affected by moderate to severe soil erosion from water in around a third of OECD member countries, although far fewer countries suffer a similar level of soil erosion from wind processes. These figures probably underestimate the number of nations affected by higher levels of soil erosion rates, as a number of key countries affected by these soil erosion processes are missing from the OECD data set, or data has not been updated for more than 20 years.

The overall trend in soil erosion across the OECD suggests one of continuing improvement over the past two decades since 1990, or at least stability in most cases, in terms of the increasing share of agricultural land affected by tolerable or lower rates rather than higher rates of erosion. Most countries have programmes that target the reduction of soil erosion risks, using a mix of payments, regulations and farm advice. These programmes include, for example, transfers of arable land to grassland, extensive use of pastures, green cover mainly in the winter period, and promoting conservation tillage practices.

Water resource withdrawals and irrigation

Managing water resources in agriculture includes: irrigation to smooth water supply across the production season; management of floods, droughts, and drainage; conservation of ecosystems; and meeting societal, cultural and recreational needs linked to water. For those regions reliant on irrigation to supplement rainfall, water is mainly drawn from surface water (rivers, lakes, reservoirs) and groundwater (shallow wells and aquifers), and only to a limited extent are recycled wastewater and desalinated water used.

Overall the key trends in total OECD agriculture's withdrawal of freshwater resources over two decades from 1990 to 2010, show that withdrawals of freshwater resources by agriculture have declined over the decade of the 2000s compared to an increase over the 1990s (Figure 0.1). Agriculture is the major user of total freshwater withdrawals, although this share varies considerably across countries (Table 1.1). Changes in the area irrigated have mirrored the trends in agricultural water withdrawals, with a slight increase over the 1990s, but decreasing over the last decade (Figure 0.1). Also the efficiency of water application on irrigated land improved for most countries over the 2000s (i.e. less water applied per hectare irrigated) compared to a more variable performance over the 1990s. These changes have mainly been driven by a mix of factors, varying between countries, including: a near stable or reduction in the area irrigated; improvements in irrigation water management and technologies; drought; release of water to meet environmental needs; and a slowdown in the growth of agricultural production.

Agriculture abstracts an increasing share of its water supplies from groundwater. Groundwater withdrawals for irrigation above recharge rates in some regions of notably, **Australia**, **Greece**, **Italy**, **Mexico** and the **United States,** is undermining the economic viability of farming in affected areas. Agriculture is also a major and growing source of groundwater pollution, especially where groundwater provides a major share of water supplies for both human needs and the farming sector.

In some OECD countries water stress is an issue which in future could have implications for freshwater withdrawals by agriculture. Water stress is based on the ratio of total freshwater withdrawals (across the economy, including agriculture) to total annual renewable freshwater availability. Countries with a medium water stress (notably Belgium, Italy, Korea, Spain), also have agricultural sectors which account for over 40% of total water withdrawals. Israel stands out as one of the world's most severely water stressed countries, with a ratio of water withdrawal to annual water availability of around 90%.

Irrigated agriculture provides a major share of the value of farm production for some OECD countries, and supports rural employment, notably in Australia, Chile, Greece, Japan, Korea, Israel, Italy, Mexico, Portugal, Spain, Turkey and the United States. In these countries agriculture accounts for over 40% of total freshwater withdrawals.

Critical to increasing agricultural production from irrigated land, improving the profitability of irrigated agriculture, and in making water savings in areas of water stress, is ***raising the physical (technical) and economic (value of output per unit of water withdrawn) productivity of water withdrawals***. This is being achieved in many OECD countries, through better water management and uptake of more efficient technologies, such as drip irrigation and lining irrigation canals. Moreover, in some cases, policy reforms have sought to transmit the cost of supplying water to irrigators by lowering support for water supplied to agriculture and lowering support for energy to pump water for irrigation. For some countries these policy reforms have led to the allocation of water to higher valued commodities which frequently require less water, such as vines and horticultural crops.

In most OECD countries, however, water policy reforms have yet to significantly reduce irrigators water application rates (megalitres of water per hectare of irrigated land). Much of the decrease in water application rates over the past 20 years has been largely driven by improvements in irrigation technologies and management practices. **Australia** stands out as the OECD country making the largest improvement in irrigation water application rates consistently over the period from 1990 to 2010, by not only improving irrigation technologies and management practices, but also undertaking policy reforms that have changed water property rights, created water trading markets, and increased water supply charges to farmers. **Israel** has also undertaken significant water policy reforms leading to, in particular, an increase in the charges paid by irrigators for water supplies, which has stimulated a reduction in water application rates per hectare irrigated and led to improvements in irrigation technologies and management.

Water quality – Nitrates, phosphorus and pesticides

Given the success across almost all OECD countries of reducing water pollution from point sources (e.g. sewage treatment, industry) focus has now switched in many countries to addressing agricultural water pollution. This is because agricultural water pollution principally originates from farms spread across the landscape (diffuse source pollution), although agriculture is also a point source of water pollution, for example, from intensive livestock farms and the disposal of residual pesticides.

The overall trends of agricultural water pollution from nitrates, phosphorus and pesticides across OECD countries are mixed over the period 2000 to 2010, but there appear few situations where significant improvements are reported. Recent national assessments of water pollution, together with limited data on national trends in agricultural water pollution, show a variable picture between countries in terms of the: trends of agricultural water pollution by contaminant type; contribution of agriculture in total pollution; and the extent to which contaminants exceed drinking water standards (Table 1.1).

For the 15-20 OECD countries that track nutrient and pesticide concentrations in surface water and groundwater, about half record that ***10% or more monitoring sites in agricultural areas have concentrations that exceed national drinking water limits***. However, the availability of measurements from monitoring sites vary greatly between countries, contaminants, surface water and groundwater. For example, the share of monitored sites where pesticide concentrations are above drinking water limits for surface and groundwater supplies are generally lower than for nutrients, but concerns remain for pesticide pollution of groundwater.

The water consumed by most of the population across OECD countries, however, is well within drinking water standards due to effective treatment to remove these pollutants, which is estimated to cost water treatment companies and consumers billions of dollars annually. But in some rural areas of OECD countries, which are not connected to treated water infrastructure systems, health concerns can be more significant from agricultural water pollution, especially where water is drawn from shallow wells.

The downward trend in nutrient surpluses and pesticide sales over the past ten years for many OECD countries, would suggest that pressure from agriculture on water systems has eased (Figure 0.1). Moreover, overall improvements in slowing rates of soil erosion on agricultural land across many OECD countries, would also indicate that the risk of agricultural water pollution could be declining, as soil sediment is a major pollutant of water systems, including the transportation by soil particles of pollutants into water.

The apparent dichotomy between decreasing agricultural pollutant loads but stable or deteriorating readings of water pollution at monitoring sites is largely explained by time lags. A time lag is the time elapsed between the adoption of new management practices by farmers and the detection of measurable improvement in water quality of the target water body. The magnitude of the time lag is highly site (surface or groundwater) and contaminant specific, and can take from hours to decades between changes in agricultural practices and improvements in water quality.

Ammonia emissions – Acidification and eutrophication

Ammonia emissions can have adverse impacts on human and animal health and on the environment. Close to the source of emission, high concentrations of ammonia may affect the respiratory system of human beings and animals, and disrupt the physiology of plants, and contribute, at longer distances from the source, to the acidification and the eutrophication of soils and water.

Overall trends in OECD agricultural ammonia emissions declined over the 2000s following a small increase over the 1990s (Figure 0.1). This conclusion needs to be qualified, as a fifth of OECD countries do not report ammonia emission trends. The reduction in agricultural ammonia emissions can be expected to have reduced the pressure on the health of humans and animals, as well as ecosystems. Agriculture is the main source of ammonia emissions accounting for 91% of emissions in 2008-10, of which over 90% is derived from livestock (Table 1.1).

The share of agricultural ammonia emissions in total acidifying emissions has risen, despite the reduction in these emissions over the 2000s. This is because emissions of other acidifying gases (sulphur dioxide, nitrogen oxides, and ammonia) from industry and the energy sector have been reduced more sharply over the past decade compared to agricultural emissions. Agricultural ammonia emissions mainly derive from livestock (manure and slurry), to a lesser extent from the application of inorganic fertilisers to crops, and also from decaying crop residues.

For most OECD countries over the past decade agricultural ammonia emissions have declined more rapidly than changes in livestock production. This apparent environmental efficiency gain in reducing the level and rate of release of agricultural ammonia emissions can be primarily attributed to the increasing numbers of farmers adopting technologies (e.g. covered manure storage facilities) and practices that are helping to reduce emissions, such as precision fertiliser application. The adoption of these technologies and practices is partly due to the use of various policy instruments in many OECD countries, for example, regulations on the storage and spreading of manure, and payments for manure storage.

As an atmospheric gas ammonia is very mobile and can move across national boundaries. In an international effort to curb ammonia and other acidifying emissions the Protocol to Abate Acidification, Eutrophication and Ground Level Ozone (the *Gothenburg Protocol*) was adopted in 1999 by some OECD countries under the *Convention on Long-rang Transboundary Air Pollution*. The *Gothenburg Protocol* set national ammonia emission ceilings for 2010, except for **Canada** and the **United States**, while the **EU** Directive on *National Emission Ceilings* has set ammonia emission ceilings at levels identical to those of the *Gothenburg Protocol*.

In terms of progress towards achieving the emission targets set for 2010 under the Gothenburg Protocol, there is a varied picture across OECD countries. By 2008-10 many countries had reduced their emissions to meet their target levels under the Protocol, but some countries will need to achieve further emission reductions to attain the 2010 target, especially Denmark and Finland. All OECD countries, however, are encouraging widespread adoption of farm nutrient management practices and implementing programmes that seek to reduce ammonia emissions.

Greenhouse gas emissions – Climate change

Linkages between agriculture and climate change are complex compared to many other economic activities because agriculture: contributes to emissions of greenhouse gases (GHGs); provides a carbon sink function under certain management practices; while agriculture is also subject to the impacts of climate change. All OECD countries are committed to GHG emission reduction, but there are no specific reduction targets set for methane or nitrous oxide, and agriculture like other sectors does not have specific commitments under the United Nations Framework Convention on Climate Change (UNFCCC). However, all OECD countries are developing agricultural climate change programmes that aim to reduce GHGs, develop carbon sinks, and make agriculture more resilient to climate change impacts.

Over the past decade total gross OECD agricultural GHG emissions decreased compared to a small increase over the 1990s, leading to an overall reduction of nearly 44 million tonnes of carbon dioxide (CO_2) equivalent over the decade 2000 to 2010 (Figure 0.1). A few countries registered an increase in GHG emissions over the 2000s, notably **Canada**, **Chile**, **New Zealand**, and the **United States** where GHG emissions increased by over 11 million tonnes of CO_2 equivalent. Over the same period reduction of agricultural GHG emissions in the **EU15** led to a saving of nearly 40 million tonnes of CO_2 equivalents. The share of agriculture in total OECD GHG emissions was relatively small in 2008-10, but averaged much higher for nitrous oxide (N_2O) and methane (CH_4) (Table 1.1).

Trends in agricultural GHG emissions are principally determined by variations in livestock production leading to changes in methane (CH_4) emissions, and crop production linked to fertiliser use affecting changes in nitrous oxide (N_2O) emissions. Relating changes in agricultural production to agricultural GHG emissions over the period 1990-2010, indicates that overall there has been an improvement in environmental efficiency of agricultural GHG emissions, that is reductions in GHG emissions have been greater than corresponding changes in agricultural production.

The environmental efficiency gains in reducing the level and rate of release of agricultural GHG emissions over the past decade, can be primarily linked to the uptake of improved technologies and farm management practices, as well as incentives to lower emissions provided by a range of policies introduced by OECD countries. These policy instruments include, for example, providing farm advice to improve livestock feed efficiency and livestock growth rates to limit GHG emissions, and payments for biodigesters to replace highly emitting sources of energy, such as coal.

Methyl bromide – Ozone depletion

Methyl bromide is used as a fumigant in the agriculture, horticulture and food sectors, but is destructive as an ozone-depleting substance which is of concern for human health and the environment. The Parties to the *Montreal Protocol on Substances that Deplete the Ozone Layer* agreed in 1997 to a global phase-out schedule for methyl bromide. Under the schedule, most OECD countries had to reduce methyl bromide use by 100% by 2005, compared to 1991 levels. Developing countries (i.e. Article 5 member countries under the *Montreal Protocol*, including among OECD countries **Chile**, **Korea**, **Mexico** and **Turkey**) started a freeze on use in 2002 at average 1995-98 levels, and needed to have achieved a 20% reduction by 2005 and 100% by 2015.

Most OECD countries have achieved the reduction level targets for methyl bromide specified under the Montreal Protocol up to 2010. Some OECD countries, however, up to 2010 were still using methyl bromide for critical uses beyond the agreed phase-out date of 2005 under the Protocol, notably the United States, and to a lesser extent Israel and Japan. This group of countries have made a significant reduction in methyl bromide use by around 90% or more by 2010, while since 2012 there has been a complete ban on methyl bromide use in Israel. For the four OECD countries – Chile, Korea, Mexico and Turkey – covered under Article 5 of the Montreal Protocol, methyl bromide use also decreased. Korea and Turkey have already eliminated methyl bromide use completely, while Chile and Mexico have made reductions in use beyond that required under the Protocol to date.

World use of total ozone depleting potential (ODP) products declined by 95% during the period 1991 to 2010, with the reduction in methyl bromide slightly less at 89%. There has also been little change in methyl bromide's share in world total ODP use at around 5% over this period (Table 1.1). OECD countries' share of world total methyl bromide use also declined over this period (Table 1.1). These reductions in methyl bromide use have been achieved by a combination of government regulations and changes in the market, as well as pressure from non-governmental organisations and the activities of private companies.

For a few OECD countries, the phase-out schedule for methyl bromide has posed a technical challenge in terms of finding alternatives, in particular, its use in the horticultural sector. In view of these difficulties the *Protocol* allows Parties to apply for **Critical Use Exemptions** (CUEs) when there are no alternatives, in addition to the existing exemption for use in quarantine and pre-shipment purposes. The CUEs are intended to give users of methyl bromide additional time to develop substitutes. In 2012, the Parties to the *Montreal Protocol* that nominated CUEs for 2012 include **Australia**, **Canada**, **Israel**, **Japan** and the **United States**. But granting CUEs may impede the effectiveness of the phase out schedule under the *Montreal Protocol* and act as a disincentive for CUE countries to seek alternatives.

Biodiversity – Farmland bird populations and agricultural land cover

Agriculture is inextricably linked to biodiversity, as agriculture produces both food and non-food commodities, and provides environmental services for society more broadly which can have, for example, scientific, recreational, and ecological value. OECD countries employ a variety of policies and approaches to reconcile the need to enhance farm production and yet reduce harmful biodiversity impacts, especially on wild species (e.g. birds) and ecosystems (e.g. semi-natural habitats). Also most OECD countries are signatories to international agreements of significance for biodiversity conservation, such as the *Convention on Biological Diversity; Convention on the Conservation of Migratory Species of Wild Animals*; and the *RAMSAR Convention* for the protection of wetlands.

No OECD country has a monitoring programme that tracks all the multi-layers of agri-biodiversity linkages. Taking into account this limitation the OECD has so far developed just two indicators that provide a broad impression of the interaction between agriculture and biodiversity. The first is the ***farmland bird index***, which is an average trend in a group of species that use farmland for nesting or breeding. Birds can act as "indicator species" providing a barometer of the health of the environment. Being close to or at the top of the food chain, birds reflect changes in ecosystems rather rapidly compared to other species.

The second indicator tracks ***changes in the permanent pasture area*** as a proxy for semi-natural habitats. Monitoring changes in the area of agricultural semi-natural habitats, can provide information on the extent of land that is subject to relatively "low intensity" farming practices, such as pasture, where usually they are managed without much mechanical disturbance and few inputs and are subject to low animal stocking densities.

Trends in OECD farmland bird populations declined continuously over the period from 1990 to 2010 for almost all countries. The decrease in farmland bird populations, however, in most cases was less pronounced over the 2000s compared to more rapid reductions over the 1990s. In describing the broad trends for OECD countries, this needs to be treated with caution as many OECD are missing from the dataset. Partial evidence for those missing countries (**Australia, Chile, Iceland, Mexico, Japan, Korea, New Zealand** and **Turkey**), however, suggest that trends in farmland bird populations are following a similar pattern to those for the countries where data exists, with probably an overall decline over the past two decades.

The slowdown in the rate of decline of farmland bird populations over the 2000s compared to the 1990s has been partly associated with (but varying between countries): efforts since the early 1990s to introduce agri-environmental schemes aimed at encouraging semi-natural land conservation on farms (e.g. field margins, buffer strips near rivers and wetlands); changes in farm management practices, such as increasing the area under conservation tillage which has increased feed supplies for birds and other wild species; and reductions in nutrient surpluses and pesticide sales (Figure 0.1) for most countries, lowering toxic effects on birds and their food supply (e.g. worms, insects).

The further intensification of agriculture and removal of natural and semi-natural habitats in some regions of the OECD, however, continues to exert ***pressure on bird populations and other flora and fauna associated with farming.*** It is also noticeable that bird species dependent on other habitats, notably forestry, have not experienced the same rate of decline as in farmland bird species.

A major share of agricultural semi-natural habitats consists of permanent pasture, which overall for OECD countries declined continuously over the period 1990 to 2010. Given the decrease in the total OECD agricultural land area since 1990, the area of permanent pasture has also been reduced, but still accounts for two-thirds of all OECD agricultural land. Much of the reduction in the permanent pasture area has been land converted to forestry, although for some countries pasture has also been converted for cultivation of arable and permanent crops.

Interpreting the consequences for farmland birds and other wildlife species of changes in the area of permanent pasture is complex. Without knowledge of the quality of the land change and its subsequent management makes it difficult to assess these developments. While the conversion of pasture to forestry, for example, can be beneficial to biodiversity, it will depend on both the quality of farmed habitat loss to forestry and also whether the forest is developed commercially or left to develop naturally.

The fragmentation of habitats on farmland is also widely reported to have a harmful impact on biodiversity. Given the magnitude of the decline in permanent pasture across most OECD countries over the past decade, however, it is likely that this has been one of the factors influencing the overall decline in farmland bird populations, and possibly other flora and fauna dependent on pasture land.

1.2. Future outlook for the environmental performance of OECD agriculture

OECD-FAO agricultural commodity projections to 2021 indicate that the expected increase in nominal and real commodity prices could provide incentives for farmers to expand production and this could heighten environmental pressures. But this may depend on the farming practices, systems and technologies adopted by agriculture, as well as the environmental sensitivity of the location where production increases occur. At the same time, production costs are projected to rise due to increases in energy, fertilisers and feed costs, as well as growing pressure on natural resources, especially land and water. Overall, with the projected increase in output prices on the one hand and rising farm input prices on the other hand, anticipated environmental outcomes could be ambiguous depending on the intensity and location of production effects.

OECD projections suggest that under a "business as usual scenario", the anticipated expansion in production could threaten recent improvements in lowering environmental pressures achieved by farmers, especially in North America, Turkey, Australia, New Zealand

and a few European countries and regions within these countries. But environmental outcomes could be ambiguous depending on the intensity and location of production changes, and also the offsetting effects of an increase in commodity prices on the one hand and rising farm input prices on the other hand.

There are a number of recent encouraging developments, however, that may help lower environmental pressure from agriculture into the future. These include, for example: improved efficiency in use of farm inputs combined with adoption of environmentally beneficial farm practices and systems; strengethened agri-environmental and environmental policies; a further shift away from production and input related farm support; and innovations in technologies and institutions along the agro-food chain that can change farmer behaviour and the actions of the agro-food industry in raising resource productivity to the benefit of the environment.

1.3. Readers guide to the OECD Compendium of Agri-environmental Indicators

Objectives and scope

This publication continues the series *Environmental Performance of Agriculture at a Glance*. The report was first published in 2008 (OECD, 2008a), together with a more detailed companion volume the *Environmental Performance of Agriculture in OECD countries since 1990* (OECD, 2008b). The series is based on previous OECD reports, *Environmental Indicators for Agriculture* (OECD, 1997; 1999; 2001). A key objective for OECD work on agriculture and the environment is to use agri-environmental indicators (AEIs) specifically as a tool to assist policy makers by:

1. describing the current state and trends of environmental conditions in agriculture that may require policy responses (i.e. establishing baseline information for policy analysis);
2. highlighting where "hot spots" or new challenges are emerging;
3. comparing trends in performance across time and between countries, especially to assist policy makers in meeting environmental targets, threshold levels and standards where these have been established by governments or international agreements;
4. developing indicators and a primary dataset that can be drawn upon for related activities, for example, the development of green growth indicators (Annex A); and
5. providing a set of indicators and database which can be drawn on for policy monitoring, evaluation and in projecting future trends.

A variant of the Driving Force-State-Response (DSR) model provides the organising framework for this report (Figure 1.1). There are a wide range of policies (e.g. agricultural and agri-environmental policies), market (commodity markets, technology) and environmental factors (e.g. soils, weather, climate change) that drive agricultural systems, practices, input use, farm outputs and the ecosystem services provided by agriculture. This in turn has implications for the state of the environment (soil, water, air, biodiversity), which then impacts on human activities, including health, social values (e.g. recreational uses), agriculture, commercial fishing, industry and urban centres. Depending on the trends in the state of the environment and impacts on human welfare this will feed back into possibly provoking a policy and/or market response.

Following the *Key Messages* and *Executive Summary*, and this *Readers Guide* chapter, ***the report has four main sections***: Chapter 2 provides a description of the policy and market drivers impacting on the environmental performance of agriculture; Chapters 3 to 13

Figure 1.1. Linkages between policies, agricultural driving forces and the state and impact of agriculture on the environment and human welfare

Policies, markets environment →	Agriculture driving forces →	State of environment →	Impacts on human welfare
• **Policies:** ➢ *Agricultural:* e.g. commodity support ➢ *Agri-environmental:* e.g. payments for riparian buffers ➢ *Environmental:* e.g. national water policy framework • **Markets:** ➢ Commodity markets, economy technology • **Environment:** ➢ Soils, weather, slope ➢ Climate change	• **Farm systems:** e.g. integrated Farming Systems (greater precisions and resource efficiency) and Organic Farming Systems • **Farm practices:** e.g. nutrient and pesticide application, tillage and irrigation practices • **Farm input use:** e.g. nutrients (nitrogen and phosphate), pesticides, water, energy) • **Farm outputs:** e.g. crops, livestock, land cover • **Farm ecosystem services:** e.g. carbon sequestration, biodiversity conservation	• **Soil erosion** • **Water quality:** Nutrients in water Nutrients and pesticides in drinking water • **Air emissions:** Ammonia Greenhouse gases Methyl bromide • **Biodiversity:** Farmland bird populations Land cover types e.g. pasture	• **Human health:** e.g. pesticides • **Social values:** e.g. impairment of rivers and lakes for recreation, fishing, visual amenity • **Agricultural and commercial fisheries:** e.g. water pollution affecting downstream users • **Industry and urban centres:** e.g. air pollution from ammonia

Note: The reader should note that the bullet points in each box are illustrative and some are interchangeable. Soil erosion, for example, is included in the "State of the Environment" list in terms of levels of soil sediment removed from agricultural land, but could also be listed as a "Driving Force" in terms of soil sediment transport into water systems.

examines agri-environmental performance in terms of specific themes: land; nutrients; pesticides; energy; soil erosion; water resources; water quality; ammonia; greenhouse gases; methyl bromide; and biodiversity; and finally, Annex A reviews recent use of the OECD agri-environmental indicators for policy monitoring and evaluation. A summary list of the AEIs in this report, and their definitions, are provided in Annex 1.A1.

Data and information sources

The main sources of data, indicator methodologies, and country information used in this report include:

1. ***OECD member countries*** exchange of data to the OECD Secretariat up to late-2012.
2. ***OECD regular work on collecting environmental data***, through the *Environmental Data Compendium* (OECD, 2008c).
3. ***Eurostat*** (the EU Statistical Office) through a process of mutual cooperation and exchange of data, to harmonise the agri-environmental indicator data set between the OECD and the European Union.
4. ***Information and data obtained from external sources*** (Annex 1.A2), including international governmental organisations, such as FAO, the Secretariats to various international environmental agreements (e.g. *Kyoto Protocol, Gothenburg Protocol* and *Montreal Protocol*), and non-governmental organisations, such as Birdlife International.
5. ***Extensive review of research literature, databases and websites***, especially to provide additional background support to the primary AEI database.

The time trends shown in the report's figures and tables are in general expressed as average annual percentage changes for a given period of time. Because of a lack of data for certain environmental indicators, and to be consistent across indicators, average annual percentage changes are equal to the geometric growth rate between two dates. In order to reduce the influence of year-to-year fluctuations, each date corresponds, as far as data are available, to a three-year average for each indicator.

Indicator caveats, limitations and assessment against the OECD Indicator Criteria

Indicator caveats and limitations

Given the complexity of calculating a wide range of indicators, across 34 OECD countries, and covering developments since 1990, it is inevitable that there are caveats and limitations when making comparisons over time and across countries, including (Annex 1.A2):

- ***Definitions and methodologies for calculating indicators*** are standardised in most cases but not all, in particular indicators for soil erosion and water quality are not fully standardised. For some indicators, such as greenhouse gas emissions (GHGs), the OECD and the UNFCCC are working toward further improvement.
- ***Data availability, quality and comparability*** are as far as possible complete, consistent and harmonised across the various indicators. But deficiencies and data gaps remain, such as the absence of data series (e.g. water quality), variability in data coverage (e.g. energy consumption), and differences related to how the data was collected (e.g. census for land use, and field surveys for farmland birds).
- ***Spatial aggregation*** of indicators is given at the national level, but for certain indicators (e.g. nutrient balances) this can mask significant variations at the regional level. Nonetheless, some examples to highlight this spatial variation are provided in the report.
- ***Trends and ranges in indicators,*** rather than absolute levels, are important for comparative purposes across countries for many indicator areas, especially as local site specific conditions can vary considerably within and across countries. Also underlying indicator calculation methodologies, coefficients and primary data are not always harmonised between countries. But absolute levels are of significance where: limits are defined by governments on the basis of scientific evidence (e.g. nitrates in water); targets agreed under national and international agreements (e.g. ammonia emissions); or where the contribution to global pollution is important (e.g. greenhouse gases).
- ***Agriculture's contribution to specific environmental impacts*** is sometimes difficult to isolate, especially for areas such as soil and water quality, where the impact of other economic activities is important (e.g. forestry) or the "natural" state of the environment itself contributes to pollutant loadings (e.g. water may contain levels of naturally occurring nitrates and phosphorus).
- ***Environmental improvement or deterioration*** is in most cases clearly revealed by the direction of change in the indicators (e.g. soil erosion, greenhouses gases), but in some cases changes can be ambiguous. Illustrative is where a farmer to meet water quality regulations reduces the nutrient content of manure spread on fields (nitrogen balance indicator) by releasing more nitrogen from stored manure into the air as ammonia (ammonia emission indicator).
- ***Baselines, threshold levels or targets for indicators*** are generally not used to assess indicator trends in the report as these may vary between countries and regions due to difference in environmental and climatic conditions, as well as national circumstances. But for some indicators threshold levels are used to assess indicator change, for example, internationally agreed targets compared against indicators trends (e.g. ammonia emissions and methyl bromide use).

Assessing the Indicators against the OECD Indicator criteria

The indicators developed by the OECD need to satisfy a set of criteria of: policy relevance; analytical soundness (scientific rigour); measurability; and ease of interpretation, to aid comparability over time and across countries. How far do the OECD agri-environmental indicators match up to these established criteria (see also the summary in Annex 1.A2)?

- ***Policy relevance*** – A key requirement of the indicators is that they adequately track developments that are of broad policy and public concern. The number of OECD countries covered by each indicator is summarised in Annex 1.A2. In most cases country coverage is representative of the OECD membership, although notably country coverage of indicators related to agriculture's impact on soil and water resources is more limited than for other indicators, partly because these are issues not of universal importance across OECD countries, unlike water quality and biodiversity for example. The contribution of the agricultural sector to specific environmental impacts, where relevant, are summarised in Table 2.1, drawing on Annex 1.A2, highlighting the importance of agriculture in the broader economy (e.g. land use, water withdrawals).
- There are some policy relevant agri-environmental indicators that are not currently available across a representative group of OECD countries. For example, indicators related to environmental farm management practices (e.g. adoption of conservation tillage, farmers undertaking soil nutrient testing, uptake of efficient irrigation water application technologies); carbon sequestration, and indicators that track pesticide risks and biodiversity conservation. As agri-environmental policies evolve, however, more countries are beginning to develop indicators across some of these areas, such as carbon sequestration in agriculture in the context of climate change mitigation policies.
- ***Analytical soundness*** – The scientific understanding of biophysical relationships and their interaction with farming activities is variable. Nutrient balances and soil erosion indicators are based on robust scientific understanding of nitrogen cycles and soil transport and fate models. But in a number of other areas there is still incomplete knowledge. For example, knowledge of the pathways and extent of agricultural pollutants into groundwater is poor. The variability in the analytical soundness of the indicators is also reflected in differences in the certainty between indicator estimates.
- ***Measurability*** – The measurability of indicators depends on robust data coverage and quality, which varies across countries. While certain data are regularly collected across most countries through agricultural census (e.g. land area), surveys are also frequently used to collect environmental data (such as water quality indicators), but country coverage is typically patchy.
- Countries often differ in the definitions of data coverage. For example, in some countries, pesticide and energy data only include agriculture, but for other countries they also cover other activities, such as forestry. Governments are usually the main institution that collect data to calculate indicators, but OECD has also drawn on other international organisations (e.g. UNFCCC for greenhouse gases) and non-governmental organisations (e.g. Birdlife International for trends in farmland bird populations).
- ***Interpretation*** – In most cases, the indicators are easy to interpret by policy makers and the wider public, but some indicators remain difficult to understand without specialist knowledge, such as indicators of biodiversity. But the interpretation of indicator results

needs to be undertaken with great care and in some cases absolute numbers cannot be compared between countries.

- OECD average trends can mask wide differences between countries, while national indicator trends can also hide large regional and local variations, as is especially the case for nutrient surpluses and water pollution. There are also marked disparities in absolute indicator levels between countries, notably nutrient surpluses, pesticide sales, energy consumption, water withdrawals, and air emissions.

The caveats to the interpretation of the indicators in this report need to be viewed in a broader context. In many cases they also apply to other indicators regularly used by policy makers. For example, there can be wide variations around national averages of socio-economic indicators (e.g. unemployment, average family income), and methodological and data deficiency problems are not uncommon (e.g. wealth distribution).

Work on establishing agri-environmental indicators is relatively recent compared with the much longer history of developing economic indicators, such as Gross Domestic Product. Measuring causal linkages between the biophysical environment and human activities through indicators is more complex than monitoring trends in socio-economic phenomena, given that many agri-environmental effects do not benefit from having market (monetary) valuations, and are not even easily measured in physical terms (e.g. farmland birds).

Structure of the Report

Following this *Overview*, Parts I and II are structured as follows:

- Chapter 2: Policy and market drivers impacting on the recent and future environmental performance of agriculture.
- Chapter 3: Agricultural production, land use, organic farming and transgenic crops.
- Chapter 4: Nutrients: Nitrogen and phosphorus balances.
- Chapter 5: Pesticides sales.
- Chapter 6: Energy: On-farm energy consumption and production of biofuels from agricultural feedstocks.
- Chapter 7: Soil: Water and wind erosion.
- Chapter 8: Water resource withdrawals, irrigated area and irrigation water application rates.
- Chapter 9: Water quality: Nitrates, phosphorus and pesticides.
- Chapter 10: Ammonia emissions: Acidification and eutrophication.
- Chapter 11: Greenhouse gas emissions: Climate change.
- Chapter 12: Methyl bromide use: Ozone depletion.
- Chapter 13: Biodiversity: Farmland bird populations and agricultural land cover.

Each of these chapters, except Chapter 2, has a common structure as follows:

- Policy context
  - The issue
  - Main challenges
- Indicators
  - Definitions

- Concepts, interpretation, limitations and links to other indicators
- Measurability and data quality
- Main Trends
- References

For the indicators in Chapters 3 to 13 (see list Annex 1.A1), all the primary data used in their calculation and the cross country time series are included on the OECD website at: *www.oecd.org/tad/sustainable-agriculture/agri-environmentalindicators.htm*.

Annex A provides a discussion of how AEIs are being used in policy analysis and monitoring. The chapter examines the use of AEIs by OECD member countries; in OECD Secretariat reports; by other International Organisations; and by the research community.

References

OECD (2008a), *Environmental Performance of Agriculture at Glance*, Paris, OECD Publishing, *www.oecd.org/tad/sustainable-agriculture/agri-environmentalindicators.htm*.

OECD (2008b), *Environmental Performance of Agriculture in OECD Countries since 1990*, OECD Publishing, *www.oecd.org/tad/sustainable-agriculture/agri-environmentalindicators.htm*.

OECD (2008c), *Environmental Data Compendium 2008*, OECD Publishing, *www.oecd.org/env*.

OECD (2001), *Environmental Indicators for Agriculture Volume 3: Methods and Results*, OECD Publishing, *www.oecd.org/tad/sustainable-agriculture/agri-environmentalindicators.htm*.

OECD (1999), *Environmental Indicators for Agriculture Volume 2: Issues and Design*, OECD Publishing, *www.oecd.org/tad/sustainable-agriculture/agri-environmentalindicators.htm*.

OECD (1997), *Environmental Indicators for Agriculture Volume 1: Concepts and Frameworks*, OECD Publishing, *www.oecd.org/tad/sustainable-agriculture/agri-environmentalindicators.htm*.

ANNEX 1.A1

Coverage of agri-environmental indicators in the OECD compendium of agri-environmental indicators

Theme	Indicator title[1]	Indicator definition[2]
I. Soil	**Soil erosion** Chapter 7	1. Agricultural land affected by water and wind erosion, classified as having moderate to severe water and wind erosion risk
II. Water	**Water resources** Chapter 8	2. Agricultural freshwater withdrawals
		3. Irrigated land area
		4. Irrigation water application rate – megalitres of water applied per hectare of irrigated land
	Water quality Chapter 9	5. Nitrate, phosphorus and pesticide pollution derived from agriculture in surface water, groundwater and marine waters
III. Air and climate change	**Ammonia** Chapter 10	6. Agricultural ammonia emissions
	Greenhouse gases Chapter 11	7. Gross total agricultural greenhouse gas emissions (methane and nitrous oxide, but excluding carbon dioxide)
	Methyl bromide Chapter 12	8. Methyl bromide use, expressed in tonnes of ozone depleting substance equivalents
IV. Biodiversity	**Farmland birds** Chapter 13	9. Populations of a selected group of breeding bird species that are dependent on agricultural land for nesting or breeding
	Agricultural land cover Chapter 13	10. Agricultural land cover types – arable crops, permanent crops and pasture areas
V. Agricultural inputs and outputs	**Production** Chapter 3	11. Agricultural production volume – index of change in total agriculture, crop and livestock production
	Nutrients Chapter 4	12. Gross agricultural nitrogen and phosphorus balances, surplus or deficit
	Pesticides Chapter 5	13. Pesticide sales, in tonnes of active ingredients
	Energy Chapter 6	14. Direct on-farm energy consumption
		15. Biofuel production to produce bioethanol and biodiesel from agricultural feedstocks
	Land Chapter 3	16. Agricultural land use area
		17. Certified organic farming area
		18. Transgenic crops area

Notes: All indicators concern primary agriculture, unless otherwise indicated in the notes of the Annex 1.A2 Table.

1. The relevant chapter of the report is indicated for each indicator title.
2. The definition of indicators are elaborated in the text for each respective chapter of the report. Broad definitions are provided here for each respective key indicator theme, but for most of the indicators listed here they are expressed as a subset of indicators. Agricultural water resources are illustrative, with the following set of indicators included under this chapter of the report:

- Trends in agricultural freshwater withdrawals, million m^3 1990 to 2010.
- Average annual growth rate of agricultural freshwater withdrawals, 1990 to 2010.
- Average annual growth rate of total freshwater withdrawals, 1990 to 2010.
- Share of agriculture freshwater withdrawals in total freshwater withdrawals, average 2008-10.
- Trends in irrigated area, hectares, 1990 to 2010.
- Average annual growth rate in irrigated area, 1990 to 2010.
- Share of irrigated area in total agricultural land area, average 2008-10.
- Share of irrigation freshwater withdrawals in total agricultural freshwater withdrawal, average 2008-10.
- Trends in irrigation water application rates, megalitres per hectare of irrigated land, 1990 to 2010.
- Average annual growth rate in irrigation water application rates, 1990 to 2010.

Trends over time for all indicators are measured in terms of annual growth rates using three year average periods as follows: 1990-92 to 1998-2000 and 1998-2000 to 2008-10. In most figures, country ranking is in the order from the highest positive growth rate to the lowest growth rate (which can be negative), for the most recent time period, i.e. 2008-10.

Where an indicator is expressed as a share of another variable this is usually over a three-year period, e.g. 2008-10. Three-year averages are important for agri-environmental indicators to help avoid the problem of extreme years, mainly associated with variable annual weather.

ANNEX 1.A2

Indicators assessed according to the OECD Indicator Criteria

	Soil erosion (water and wind)		Water resources		
	1. Share of agricultural land classified as having moderate to severe water erosion risk	1. Share of agricultural land classified as having moderate to severe wind erosion risk	2. Share of agricultural freshwater withdrawals in total freshwater withdrawals	3. Share of irrigated land area in total agricultural land	4. Irrigation water application rates
Policy relevance					
1. Number of countries[1]	27	18	29	26	15
2. Contribution of agriculture to environmental impact[2]	n.c.	n.c.	44%	4%	n.c.
Analytically sound					
1. Science of calculation methodology[3]	Sound	Sound	Average	Sound	Average
2. Certainty of indicator estimate[4]	Low	Low	Average	High	Average
Measurable					
1. Period covered by indicator[5]	1990-2010	1995-2010	1990-2010	1990-2010	1990-2010
2. Frequency of data collection[6]	Infrequent	Infrequent	Annual	Annual	Annual
3. Method of primary data collection[7]	Survey/model	Survey/model	Field/model	Census	Field/model/census
4. Data coverage[8]	Agriculture	Agriculture	Agriculture	Agriculture	Agriculture
5. Institution collecting data[9]	Government	Government	Government	Government	Government
Interpretation					
1. Easy to interpret[10]	High	High	High	High	High
2. Cross-country comparability[11]	Yes	Yes	Yes	Yes	Yes

		Water quality							
		5. Share of agriculture nutrients in:		5. Share of monitoring sites in agricultural areas that exceed drinking water limits for nutrients in:		5. Share of monitoring sites in agriculture areas that exceed drinking water limits for pesticides in:		5. Share of monitoring sites in agricultural areas where one or more pesticides are present in:	
Policy relevance									
1. Number of countries[1]		Surface water	Coastal water	Surface water	Ground water	Surface water	Ground water	Surface water	Ground water
	N:	13	9	15	21	11	14	7	6
	P:	12	8	7	..				
2. Contribution of agriculture to environmental impact[2]	N:	n.c.	n.c.	n.c.	n.c.	n.c.	n.c.	n.c.	n.c.
	P:	n.c.	n.c.	n.c.	n.c.	n.c.	n.c.	n.c.	n.c.
Analytically sound									
1. Science of calculation methodology[3]		Sound		Sound		Average		Average	
2. Certainty of indicator estimate[4]		High		High		Average		Average	
Measurable									
1. Period covered by indicator[5]	N:	2000-09	2000-08	2000-10	2000-10	2000-10		2000-10	
	P:	2000-09	2000-08	2000-10	..	2000-10		2000-10	
2. Frequency of data collection[6]		Infrequent		Infrequent		Infrequent		Infrequent	
3. Method of primary data collection[7]		Sample survey		Sample survey		Sample survey		Sample survey	
4. Data coverage[8]		Agriculture		Agriculture		Agriculture		Agriculture	
5. Institution collecting data[9]		Government		Government		Government		Government	
Interpretation									
1. Easy to interpret[10]		High		High		High		High	
2. Cross-country comparability[11]		Yes		Yes		Yes		Yes	

	Ammonia	GHGs	Methyl bromide	Wild species diversity
	6. Share of agricultural ammonia emissions in total ammonia emissions	7. Share of agricultural GHG emissions in total OECD GHG emissions	8. Share of agricultural methyl bromide use in world total methyl bromide use	9. Trends in selected species of farmland birds that are dependent on agricultural land for breeding and nesting
Policy relevance				
1. Number of countries[1]	27	34	34	20
2. Contribution of agriculture to environmental impacts[2]	91%	8%	46%	n.c.
Analytically sound				
1. Science of calculation methodology[3]	Sound	Sound	Sound	Sound
2. Certainty of indicator estimate[4]	High	High	High	High
Measurable				
1. Period covered by indicator[5]	1990-2010	1990-2010	1991-2010	1990-2010
2. Frequency of data collection[6]	Annual	Annual	Annual	Annual
3. Method of primary data collection[7]	Model	Model	Model	Survey
4. Data coverage[8]	Agriculture	Agriculture, forestry, fisheries	Agriculture + agro-food sector	Agriculture
5. Institution collecting data[9]	UNECE	UNFCCC	UNEP	National BirdLife organisations and BirdLife International (NGO)
Interpretation				
1. Easy to interpret[10]	Average	High	Average	High
2. Cross-country comparability[11]	Yes	Yes	Yes	Yes

	Nutrients		Pesticides	Energy	Biofuel	
	12. Gross nitrogen balance	12. Gross phosphorus balance	13. Pesticide sales in tonnes of active ingredients	14. Direct on-farm energy consumption	15. Bioethanol production	15. Biodiesel production
Policy relevance						
1. Number of countries[1]	33	33	33	33	15	14
2. Contribution of agriculture to environmental impact[2]	n.c.	n.c.	n.c.	1.6%	n.c.	n.c.
Analytically sound						
1. Science of calculation methodology[3]	Sound	Sound	Average	Average	Sound	Sound
2. Certainty of indicator estimate[4]	High	High	Average	Average	High	High
Measurable						
1. Period covered by indicator[5]	1990-2009	1990-2009	1990-2010	1990-2010	1994-2011	2000-10
2. Frequency of data collection[6]	Annual	Annual	Annual	Annual	Annual	Annual
3. Method of primary data collection[7]	Census and coefficients	Census and coefficients	Survey	Census	Survey	Survey
4. Data coverage (i.e. agriculture + forestry + fisheries)[8]	Agriculture	Agriculture	Agriculture, forestry	Agriculture, forestry, fisheries	Agriculture, forestry	Agriculture, forestry
5. Institution collecting data[9]	Government	Government	Government	Government	Government industry	
Interpretation						
1. Easy to interpret[10]	Average	Average	High	Average	Average	Average
2. Cross-country comparability[11]	Yes	Yes	Yes	Yes	Yes	Yes

	Production	Land			
	11. Index of total agricultural crop and livestock production volume	16. Share of agricultural land in total agricultural land area	10. Share of permanent crops, arable and pasture, in total agricultural land area	17. Share of land under certified organic farming in total agricultural land area	18. Share of transgenic crops in total area, arable and permanent crop area
Policy relevance					
1. Number of countries[1]	34	34	34	34	34
2. Contribution of agriculture to environmental impact[2]	n.c.	36%	n.c.	1.9%	18%
Analytically sound					
1. Science of calculation methodology[3]	Sound	Sound	Sound	Sound	Sound
2. Certainty of indicator estimate[4]	High	High	High	High	High
Measurable					
1. Period covered by indicator[5]	1990-2010	1990-2010	1990-2010	2002-10	1996-2011
2. Frequency of data collection[6]	Annual	Annual	Annual	Annual	Annual
3. Method of primary data collection[7]	Census	Census	Census	Survey	Survey
4. Data coverage[8]	Agriculture	Agriculture	Agriculture	Agriculture	Agriculture
5. Institution collecting data[9]	FAO	Government	Government	Government + IFOAM	ISAAA
Interpretation					
1. Easy to interpret[10]	High	High	High	High	High
2. Cross-country comparability[11]	Yes	Yes	Yes	Yes	Yes

..: not available; n.c.: not calculated..

Notes: The indicators included in this annex are those listed in Annex 1.A1 using the same numbering (please note some subset indicators are included which have the same number, for example, soil erosion from water and wind), which are assessed according to the OECD indicator criteria of: policy relevance; analytical soundness; measurability; and ease of interpretation.

3. "*Number of countries*": The number of countries for which data are available for each respective indicator (e.g. 33 of the 34 OECD member countries provided data in the report on agricultural nutrient balances).
4. "*Contribution of agriculture to environmental impact*": The contribution of OECD agriculture to respective environmental impacts, where relevant, for example, OECD agriculture freshwater withdrawals accounted for 44% of total freshwater withdrawals in 2008-10. These data are summarised in Table 2.1.
5. "*Science of calculation methodology*": A qualitative assessment – sound, average, weak – of the scientific rigour of each respective indicator's method of calculation. There is no indication of the number of countries that are using the relevant methodology for each respective indicator, but in general the methodology used to calculate the indicator is the same for all countries (exceptions are clarified in the respective chapters of the report).
6. "*Certainty of indicator estimate*": A qualitative assessment – high, average, low – of the certainty of the estimate made for each indicator.
7. "*Period covered by indicator*": The period covered by the indicator is shown for each respective indicator, but is usually the period 1990 to 2010. This does not imply that for every indicator annual time series are available, for example soil erosion and water quality, as shown in the next point.
8. "*Frequency of data collection*": The frequency (e.g. annual to every five years) of primary data collection used in the calculation of the indicators.
9. "*Method of primary data collection*": The method (e.g. survey, census) used to collect the primary data to calculate the indicators.
10. "*Data coverage:* For most indicators, the data coverage is for primary agriculture, but in some cases (e.g. energy consumption and pesticide sales) data may cover other users (e.g. forestry) where countries are unable to disaggregate the data.
11. "*Institution collecting data*": The main institution with primary responsibility for collecting data to calculate the indicators.
12. "*Easy to interpret*": A qualitative assessment – high, average, low – of the ease of interpreting the indicators by policy makers and the wider public.
13. "*Cross-country comparability*": Identification (yes or no) of whether the indicators are comparable across countries. It should be noted that "Yes" may only indicate that the overall trends are comparable but not the absolute levels, for example irrigation water application rates.

Chapter 2

Policy and market drivers impacting on the recent and future environmental performance of agriculture

This chapter provides an overview of the role of agriculture in the economy and the environment, underlying the significant position of agriculture with respect to the environment. It also examines policy and market drivers affecting recent trends in the environmental performance of agriculture, such as the changes in the overall level and composition of support to farmers, developments in agri-environmental policies and trends in agricultural commodity prices. Finally, the chapter presents an outlook for the environmental performances of agriculture in relation to projected changes in agricultural commodity prices and production, and identifies developments that may help lower the pressure of agriculture on the environment and encourage the development of environmental benefits associated with agriculture.

The environmental performance of agriculture is shaped by a number of key drivers including policies, markets, technologies, farm management practices, as well as environmental conditions (e.g. soils, weather) (Figure 1.1). The use of inputs by farmers, such as fertilisers, pesticides, land and water, ultimately depend on the relative prices of agricultural outputs, inputs and farm management skills. The incentives to adopt environmentally beneficial farming practices also depend on the level and composition of agricultural producer support, overall market forces and available technologies.

2.1. Context: The role of agriculture in the economy and the environment

The role of the primary agricultural sector in the Gross Domestic Product (GDP) of OECD countries remains relatively small in most cases, although is more significant when considering the whole agro-food chain (Figure 2.1). OECD countries, however, still contribute a significant share of world agricultural production and exports for a set of commodities, such as wheat, milk, and meat. However, these shares are projected to decline over the coming decade, with the continued expansion of the industrial and service sectors in some emerging countries (OECD, 2012a).

Figure 2.1. **Gross Domestic Product structure for agriculture, OECD countries, 2009**

Share of GDP (%)

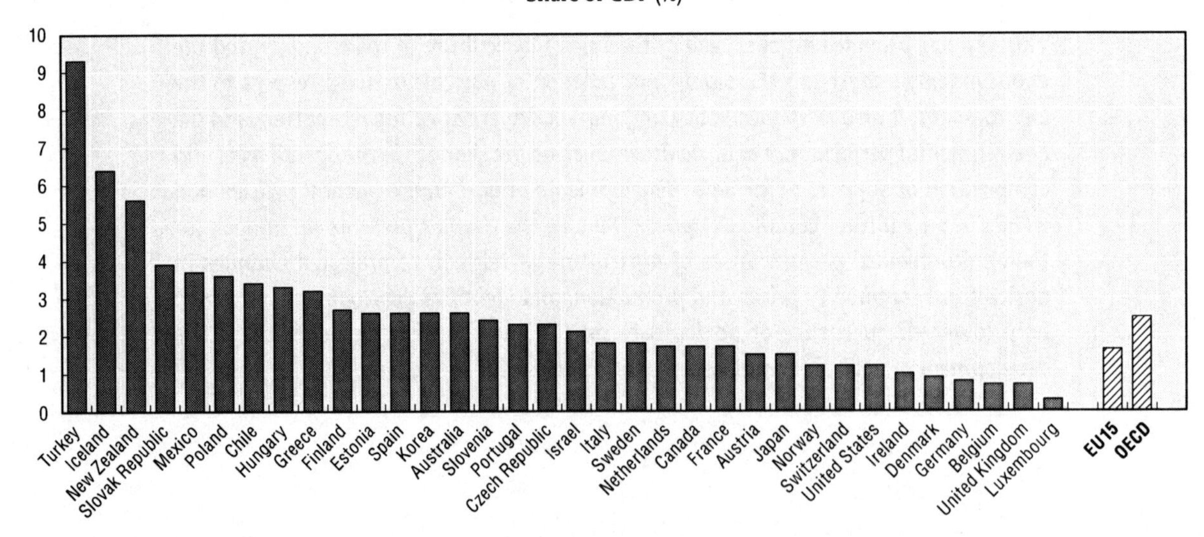

Notes: Countries are ranked from highest to lowest share of agriculture in GDP.

The OECD includes 34 OECD countries. GDP structure: agriculture includes also hunting, forestry and fishing.

The statistical data for Israel are supplied by and under the responsibility of the relevant Israeli authorities. The use of such data by the OECD is without prejudice to the status of the Golan Heights, East Jerusalem and Israeli settlements in the West Bank under the terms of international law.

Source: OECD (2011), *Towards Green Growth: Monitoring Progress: OECD Indicators*, *www.oecd.org/greengrowth*.

StatLink http://dx.doi.org/10.1787/888932792350

New patterns of agricultural commodity trade are thus expected to emerge; which could affect the extent and distribution of environmental pressures across the world, for both agricultural exporting and importing countries. This tendency could be reinforced in a context where markets become major drivers of farmers' production and investment decisions, especially if the level of support continues its recent downward trend and the composition of agricultural producer support is increasingly decoupled from production and input use (OECD, 2012b).

OECD primary agriculture has a significant position with respect to the environment, in contrast to its much smaller role in terms of its contribution to the overall economy. Agriculture produces a broad set of both positive (e.g. biodiversity conservation) and negative environmental externalities (e.g. air pollution) that are not reflected by its contribution to GDP, as usually there are no markets for these externalities. The relative importance of OECD agriculture in its use of natural resources and contribution to environmental pressures, drawing from Chapters 3 to 12 of this report, are summarised in Table 2.1.

Table 2.1. **The role of primary agriculture in the economy and the environment, OECD countries, 2008-10**

Percentage of OECD primary agriculture in total	OECD average	Range of values (minimum to maximum)
• GDP	2.6%	0.3 to 9.2%
• Land area	36%	3 to 72%
• Certified organic farm area as a share of total agricultural area	1.9%	0.01 to 15.6%
• Nutrient balances (surpluses and deficits):		
Nitrogen, kg per hectare of agricultural land	63 kg/ha	1 to 228 kg/ha
Phosphorus, kg per hectare of agricultural land	6 kg/ha	-10 to 49 kg/ha
• Pesticide sales	70%	65 to 80%
• Energy consumption	1.6%	0.4 to 6.3%
• Water withdrawals	44%	0.2 to 89%
• Irrigated land area share in total agricultural area	4%	0.4 to 54%
❖ Water pollutants, *of which:*		
Nitrates in surface water	..	33 to 82%
Nitrates in groundwater[1]	..	1 to 34%
Nitrates in coastal water	..	35 to 78%
Phosphorus in surface water	..	17 to 70%
Phosphorus in coastal water	..	23 to 50%
Pesticides in surface water[1]	..	0 to 75%
Pesticides in groundwater[1]	..	0 to 25%
• Ammonia emissions	91%	82 to 98%
• Greenhouse gas emissions	8%	2 to 46%
Of which: Nitrous oxide emissions	75%	..
Methane emissions	38%	..
• Share of OECD methyl bromide use in world total:		
Ozone depleting products	5%	..
Methyl bromide use	46%	..

..: not available.

Notes: The data in this table should be interpreted as approximate values rather than precise values, and for some indicators include forestry and fisheries. For full notes and sources, consult the website below.

1. Share of monitoring sites exceeding recommended drinking water threshold limits.

Source: OECD *Agri-environmental Indicator Database, www.oecd.org/tad/sustainable-agriculture/agri-environmentalindicators.htm.*

StatLink http://dx.doi.org/10.1787/888932793414

The main policy challenge is to progressively decrease the negative impacts and increase the positive environmental benefits associated with agricultural production so that ecosystem functions can be maintained and food security ensured for the world's growing population. This implies improving the productivity and sustainability of agro-food systems, for example, by: enhancing land management practices; minimising water and air pollution discharges from agriculture; curtailing the rate of biodiversity loss on farmland; and addressing agricultural support policies linked to production and use of inputs, that can encourage the intensity of production beyond that which would occur in the absence of these policies.

2.2. Policy and market drivers affecting recent trends in the environmental performance of agriculture

Reform in agricultural support policies across most OECD countries since 1990 have had an influence in lowering the overall pressure on the environment and encouraging environmental benefits, than would otherwise have been the case in the absence of these policy reforms, including (OECD, 2012c):

1. ***Reduction in the overall level and composition of support to farmers***. In 2009-11, support to producers in OECD countries was estimated at almost USD 250 billion (around EUR 180 billion), as measured by the Producer Support Estimate (PSE) (OECD, 2012b). The PSE fell from 37% of farmers' total receipts in 1986-88 on average to 20% in 2009-11, to a large extent due to lowering border protection and budgetary support to agriculture (Figure 2.2).

Figure 2.2. **Agricultural support and the composition of support, OECD countries, 1986-2011**

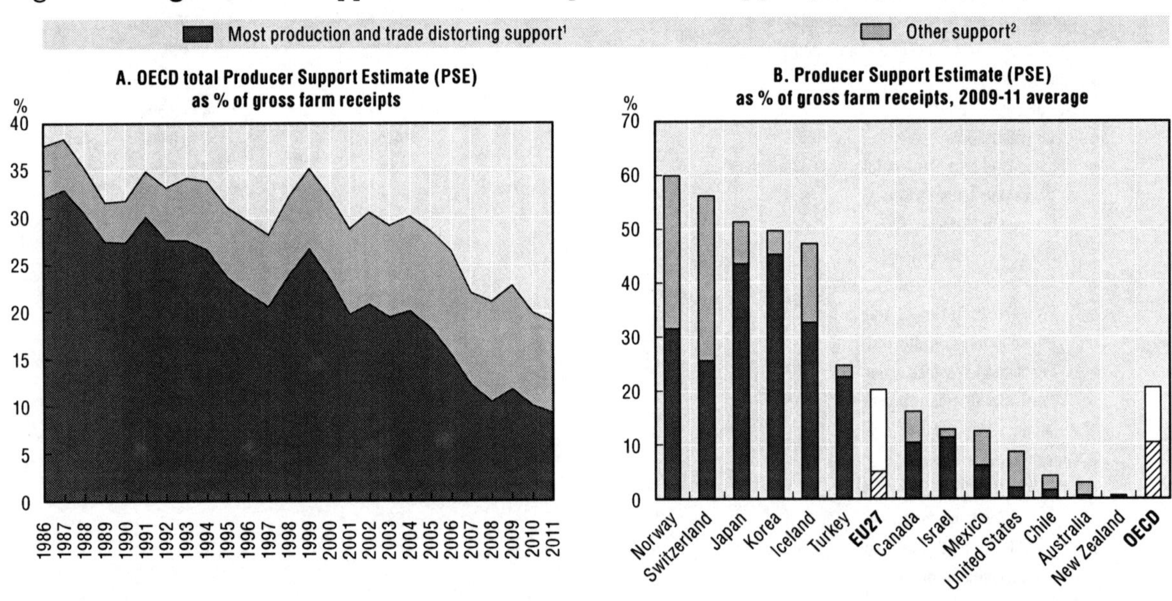

Notes: Producer Support Estimate (PSE): The annual monetary value of gross transfers from consumers and taxpayers to agricultural producers, measured at the farm gate level, arising from policy measures that support agriculture, regardless of their nature, objectives or impacts on farm production or income.

The statistical data for Israel are supplied by and under the responsibility of the relevant Israeli authorities. The use of such data by the OECD is without prejudice to the status of the Golan Heights, East Jerusalem and Israeli settlements in the West Bank under the terms of international law.

1. Most production and trade most distorting support is defined to include market price support, payments based on output and variable input use without input constraints.
2. Other support is the difference between total producer support and the potentially most distorting support.

Source: OECD, *PSE/CSE Database*, 2012, *www.oecd.org/agriculture/pse*.

StatLink http://dx.doi.org/10.1787/888932792369

Policies that increase producer prices or subsidise input use (e.g. pesticides, fertilisers, water) without restricting output encourage farmers to increase production, use more inputs, and farm more fragile lands. The opportunity costs of improving the environment in agriculture, are higher than they need be while agricultural production and input support remains. Production and input support policies by providing homogenous incentives across agriculture, fail to recognise the biophysical heterogeneity of agricultural landscapes, leading to a mismatch between the intrinsic capacity of the environment to absorb pollution and the intensity of agricultural production.

This leads to pollution hotspots where inappropriate land use and management is practised in environmentally sensitive landscapes. Agricultural commodity support can also act as a disincentive for farmers to participate in voluntary land and water conservation programmes (National Research Council, 2008). Rising commodity market prices, partly due to agricultural policy reform, may also provide a disincentive for farmers to participate in these programmes.

Policies that seek to reduce the environmental impact of farming and to improve food security also need to be well targeted to be effective. Support provided to farmers needs to encourage greater on-farm productivity and resource use efficiency to achieve environmental benefits. This combined with measures to discourage farming on fragile lands may lead to greater conservation by providing incentives for sustainable agriculture. Indeed, a key part of agricultural policy reforms in many countries is to provide incentives to farmers to develop environmentally beneficial practices that can, for example, help to control water and soil sediment flows from farmland, offer biodiversity conservation possibilities, and develop agriculture's role in carbon sequestration.

2. ***Change in the way support is delivered toward support more decoupled from production.*** The ways in which support is provided to farmers have also changed (Figures 2.2 and 2.3). OECD governments are gradually shifting to support that is more decoupled from current production and which gives greater freedom to farmers in their production choices, such as area payments. This shift in support has also led to the development of a set of targeted agri-environmental measures to reduce environmental pressures, such as regulatory requirements, payments based on land retirement or farming practices, and technical assistance. Even with more decoupled forms of support, however, such as arable crop area payments which are not environmentally neutral, this may provide incentives for bringing additional land into cultivation or to continue cultivation of marginal lands, and hence, contribute to overall environmental pressure (see Chapter 4 in OECD, 2010a).

   These measures mandate or provide incentives for farmers to adopt more environmentally beneficial farming practices, for example, the promotion of extensive farm systems and adoption of crop diversification and conservation tillage practices. The relative importance of these different types of measures varies across OECD countries. Although regulatory requirements constitute the core of these measures in OECD countries, there is a trend since the mid-1990s towards an increase of agri-environmental payments in some OECD countries.

3. ***Development of environmental conditionality.*** Support is also becoming more tied to certain conditions, as well as decoupled from production and input use. In 2006-08, over 30% of support to OECD farmers had some such conditions attached to it, whereas in 1986-88 this share was only 4% (OECD, 2010b). Increasing use of environmental conditionality (cross compliance) that links the provision (withdrawal) of support

Figure 2.3. **Level and composition of agricultural producer support, OECD countries, 1995-2011**

Direction of change, 1995-97 to 2009-11

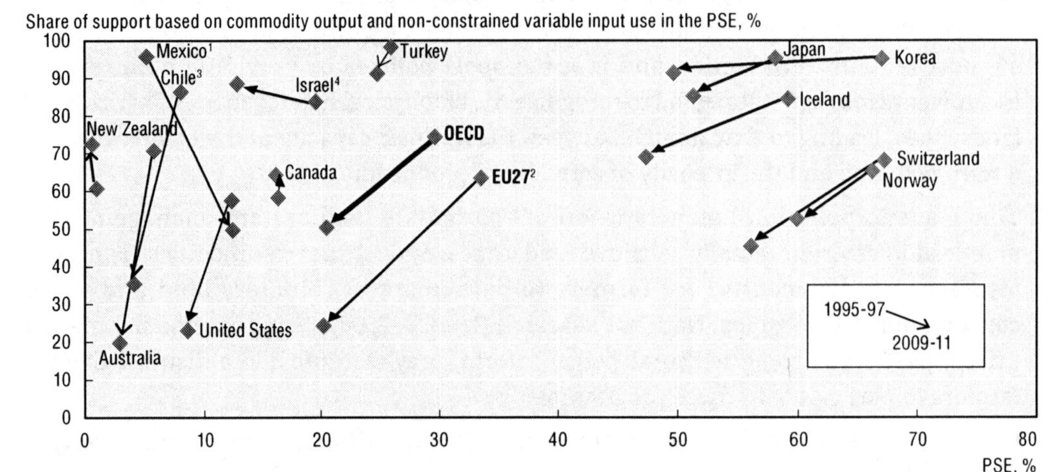

Notes: Producer Support Estimate (PSE): The annual monetary value of gross transfers from consumers and taxpayers to agricultural producers, measured at the arising farm gate level, from policy measures that support agriculture, regardless of their nature, objectives or impacts on farm production or income.

The level of support is presented by the percentage PSE. The composition of support is presented by the share in gross farm receipts of the most production and trade distorting support, including Market Price Support, Payments based on output and Payments based on non-constrained variable input use.

1. For Mexico, the change is measured between 1996-98 and 2009-11.
2. EU15 for 1995-2003; EU25 for 2004-06 and EU27 from 2007.
3. For Chile, change is measured between 1997-99 and 2009-11.
4. For Israel, change is measured between 1997-99 and 2009-11. The statistical data for Israel are supplied by and under the responsibility of the relevant Israeli authorities. The use of such data by the OECD is without prejudice to the status of the Golan Heights, East Jerusalem and Israeli settlements in the West Bank under the terms of international law.

Source: OECD, *PSE/CSE Database*, 2012, *www.oecd.org/agriculture/pse*.

StatLink http://dx.doi.org/10.1787/888932792407

payments to the requirement they meet a number of specified conditions related to their environmental performance, is being used toward addressing a wide number of environmental concerns in agriculture.

Overall across OECD countries, considering the combination of more stringent environmental regulations, increases in agri-environmental payments, and development of other measures such as market-based instruments, collective action and technical assistance, there has been a trend towards a better integration of environmental issues in farmers' decision making since the early 1990s. This is an important development in understanding the trends in agri-environmental indicators discussed in Chapters 3 to 12 of this report.

The overall decrease in agricultural producer support, in particular their most distortive components, has the natural counterpart that market prices tend to become more important as key drivers of farmers' choices. Over recent years international agricultural commodity markets have been strongly marked by higher and more volatile agricultural commodity prices. Rising real agricultural commodity prices can provide incentives to farmer to increase the scale and intensity of their production, including increasing consumption of inputs such as fertilisers, pesticides, energy and water for irrigation, between inputs and outputs, although these relationships are complex. This potentially affects the opportunity cost of adopting environmentally beneficial farming practices.

The effects of price volatility and production risks on environmental performance are much more difficult to characterise than the effect of price levels. Furthermore, the recent period of commodity price volatility occurred over a relatively short period, making it difficult to provide a robust evaluation of their consequences for agriculture and the environment. There are two countervailing effects of price volatility and production risks on the environment in agricultural systems typically found in OECD countries. On the one hand, an increase in price volatility could reduce the optimal scale of production, and hence, input use, due to farmers' risk aversion (the scale effect). On the other hand, if price volatility mainly results from production shocks due to unfavourable conditions (e.g. drought and pests), there is an incentive for farmers to increase the use of risk-reducing inputs such as irrigation water and pesticides, which could have significant consequences for the environmental performance of agriculture.

Illustrative of these developments has been the influence of changing world market conditions on the dairy industry in **New Zealand**, and the consequences for the environment, more specially nitrate pollution of water systems (Figure 2.4). Between 1990 and 2010 the New Zealand national nitrogen surplus (defined in Chapter 4 of this report), increased at a very similar annual rate to that for the national dairy cattle herd, which has been the main source of nitrogen surplus (i.e. farm manure and slurry) in New Zealand (Figure 2.4). At the same time, the profitability of the New Zealand dairy industry has benefitted from the rise in the international price of milk over this period, given there is no support or protection of the New Zealand dairy sector (this price is used as proxy for international dairy product prices, see definitions in the OECD *PSE/CSE Database*).

Figure 2.4. **World milk price, dairy cattle numbers, milk production and nitrogen surplus, New Zealand, 1990-2010**

Index 1990-92 = 100

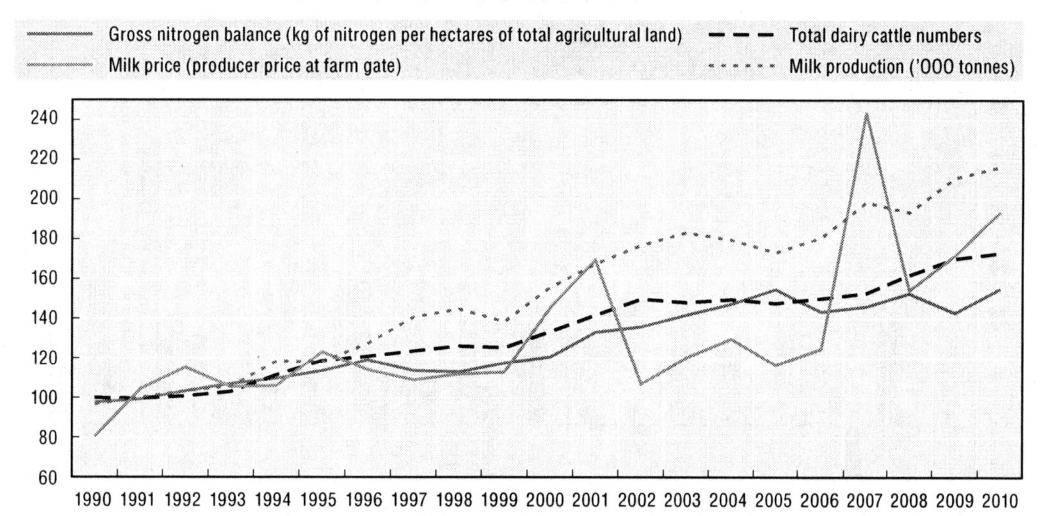

Notes: The gross nitrogen balance calculates the difference between the nitrogen inputs entering a farming system (i.e. mainly livestock manure and fertilisers) and the nitrogen outputs leaving the system (i.e. the uptake of nitrogen for crop and pasture production).

The milk price used as a proxy for the world market price, measures the transfers from consumers and taxpayers to agricultural producers arising from policy measures that create a gap between domestic market prices and border prices of milk, measured at the farm gate level.

Source: OECD/Eurostat Agri-Environmental Indicators Database; OECD *PSE/CSE Database*, *www.oecd.org/agriculture/pse*; OECD *Aglink Database*, *www.agri-outlook.org*.

StatLink http://dx.doi.org/10.1787/888932792426

The rise in the world dairy commodity prices over the past decades, but especially since the mid-2000s, has provided a considerable incentive to **New Zealand** livestock producers to intensify dairy production compared to other livestock sectors (e.g. beef and sheep). These developments present a major challenge for New Zealand policy makers and the agriculture sector. In brief, that challenge involves achieving a sustainable dairy industry responding to market signals that can capture the economic and social benefits for farmers and the wider rural community induced by higher dairy prices, while minimising the environmental pollution of rivers, lakes and groundwater from excess nutrients, as well as reducing other environmental impacts associated with dairying (e.g. diminishing greenhouse gas emissions, especially methane).

2.3. Future outlook for the environmental performance of agriculture

According to the OECD-FAO *Agricultural Outlook 2012-2021* (OECD, 2012a), in the next decade, agricultural commodity prices in nominal and real terms are likely to be higher and more volatile on average than they were in the last decade (Figure 2.5). This rise in prices would result from growing world demand for food, in relation to rising population and incomes, particularly in emerging countries, an increase in the demand for meat, and the development of biofuels. Commodity prices increases could provide incentives for farmers to boost production and this may heighten environmental pressures, depending on the farming practices, systems and technologies adopted by the sector, as well as the environmental sensitivity of the location where production increases occur.

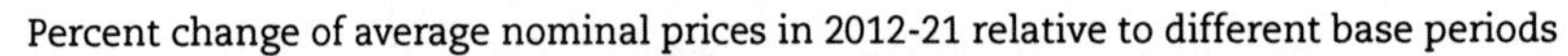

Figure 2.5. **Nominal world agricultural commodity price projections for 2012-21 relative to 2009-11 and 2002-11**

Percent change of average nominal prices in 2012-21 relative to different base periods

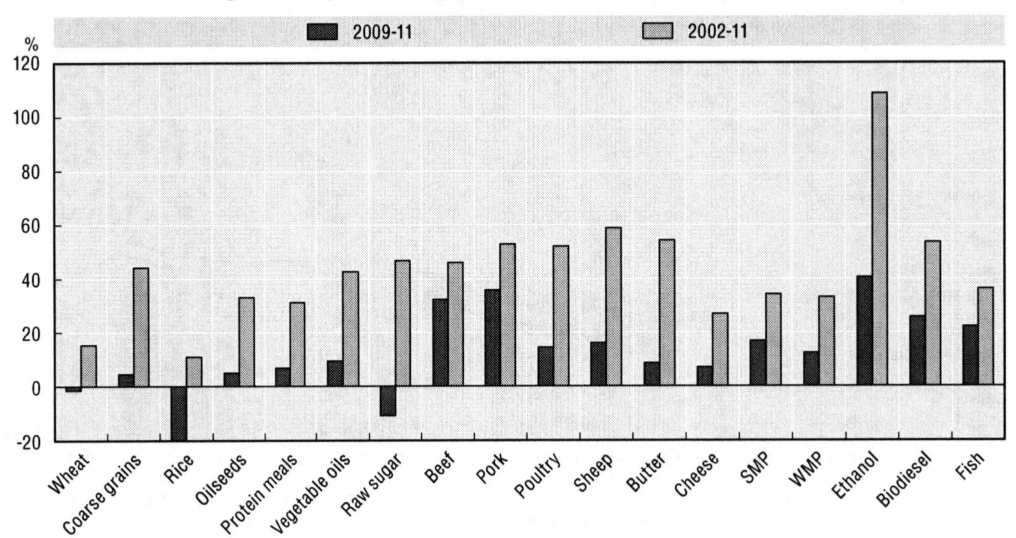

Note: SMP: Skim Milk Powder; WMP: Whole Milk Powder.

Source: OECD (2012), *OECD-FAO Agricultural Outlook 2012-2021*, *www.agri-outlook.org*.

StatLink http://dx.doi.org/10.1787/888932792445

At the same time, production costs are projected to reach higher levels than in the previous decade, due to increases in energy, fertilisers and feed costs, as well as growing pressure on natural resources, especially land and water. Over the next decade, the crude oil price is projected to rise, which would translate into higher farm input prices (e.g. fertilisers, energy to pump water, pesticides), although developments such as shale gas production in some countries could lower natural gas prices and reduce costs of nitrogen fertiliser production. Overall, with the increase in output prices on the one hand, and rising farm input prices on the other hand, the expected environmental outcomes could be ambiguous depending on the intensity and location of production effects.

With the projected increase in commodity prices, agricultural production is projected to expand over the next decade, but at a slower rate than in the preceding one, down from 1.5% to 1.2% per annum (OECD, 2012a), with significant international differences across countries and commodities. The overall reduction in the growth rate of farm output is expected to originate from slower rates of improvement in crop productivity compared to earlier decades, while cropland area is expected to remain relatively constant. The livestock sector, however, is expected to grow at a similar rate as in the previous decade.

The outlook for agricultural commodity prices translates into projected growth in agricultural production for nearly all OECD countries over the coming decade (Figure 2.6). From the trends in national agricultural production projections in Figure 2.6, it is possible to discern two broad groupings of OECD countries in terms of their potential pressure on the environment over the coming decade:

- **Group 1:** Countries which are projected to continue with strong growth in production over the coming decade, including: **Australia, Canada, Mexico, New Zealand, Turkey and the**

Figure 2.6. **Agricultural production volume projections, OECD countries, 2000-21**

Index 2004-06 = 100

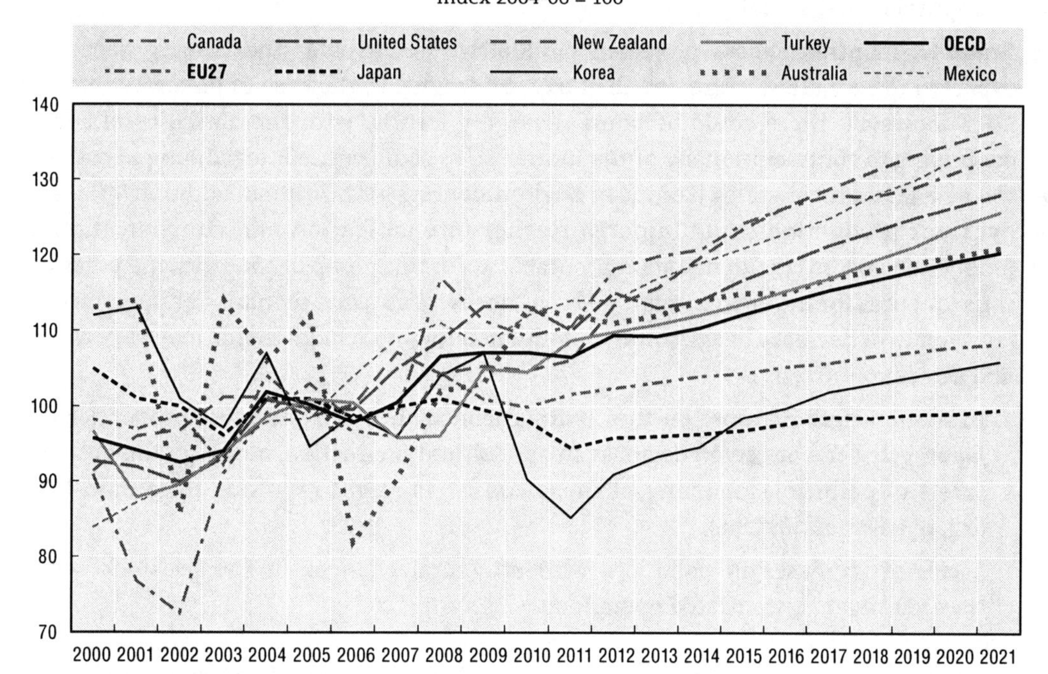

Note: Net agricultural production measures gross value of product produced, net of "internal" feed and seed inputs to avoid double counting (for example, maize and livestock production), so that the production measure approximates a value added concept.

Source: OECD (2012), *OECD-FAO Agricultural Outlook 2012-2021*, *www.agri-outlook.org*.

StatLink http://dx.doi.org/10.1787/888932792464

United States.* Most OECD countries in this group have over the past decade largely expanded production by raising productivity and intensifying production on a reduced land area. However, in regions within some of these countries there is a risk of expanding production onto environmentally fragile land or marginal land not previously cultivated. For this group of countries the potential consequences for the environmental performance of agriculture of the projected growth in agricultural production under the "business as usual scenario", might include (trends may vary within and across countries):

1. Heightened pressure on the environment from the increased use of farm inputs (e.g. fertilisers, pesticides, energy and water) and livestock (e.g. more manure, ammonia and methane emissions), although absolute levels of pollution for many of these countries are mostly below the OECD average (e.g. nutrient surplus/ha).
2. Elevated soil erosion as a result of farming more intensively productive agriculture land and/or expanding production onto marginal and fragile land and susceptible to erosion.
3. Expanded production of bioenergy which depending on the crop mix and farm management practices to produce bioenergy feedstocks may lead to heightened soil erosion and water pollution risks, especially where cereals, oilseeds and sugar crops are used as feedstocks for manufacturing biofuels.
4. Regionalised pressures on the environment could alter as a result of the continued structural changes in livestock production toward larger and more concentrated livestock operations, notably in the pig, poultry and dairy sectors, although in some cases larger, concentrated livestock operations can provide efficient levels of waste disposal management.

- **Group 2:** Countries where projected production growth over the coming decade is expected to be modest, comprise the **EU27**, or decline, in the case of **Japan**. Within the EU27, however, there could be some diverging trends, with the agricultural sector continuing to contract in many of the former EU15 countries, but expanding in some of the new EU member states (European Environment Agency, 2010). In addition, crop and livestock production could undergo further intensification and concentration of production on less land to maintain profitability. For this group of countries the potential consequences for the environment under a "business as usual scenario" of the projected low growth or decrease in agricultural production might include (trends may vary within and across countries):

1. Reduced overall pressure on the environment, with this trend more pronounced in **Japan** given the projected decrease in agricultural production, although the absolute levels of pollution for many of these countries might continue to remain high (e.g. nutrient surplus/ha);
2. Localised increases in pollution, with structural changes in the livestock sector towards larger concentrated operations.

* **Korea** is the exception in this group, with production declining from the late 1990s to present but then projected to expand back to the levels of the late 1990s, largely explained by growth in beef production stimulated by a rise in Korean consumer demand and higher government support to producers.

For all OECD countries over the medium term there are a number of developments that may generally help toward lowering the pressure of agriculture on the environment and encourage the development of environmental benefits linked to agriculture, including:

1. Efficiencies in lowering farm input use per unit of output, induced by a number of factors including for example, a changing regulatory environment leading to more targeted pesticide use; and the higher prices for inorganic fertilisers and pesticides due to the projected increase in fossil fuel (e.g. gas, oil, coal) prices, which might also encourage greater use of livestock waste as a bioenergy feedstock.
2. Improvements in farm management practices (e.g. conservation tillage), and precision agricultural technologies, such as the use of on-farm global positioning systems (GPS), that can lead to more efficient use of inputs, and also innovations in the agro-food industry (e.g. inputs, seeds and production processes) that could bring benefits by increasing resource efficiency on-farms and lowering environmental pressures along the whole agro-food chain.
3. Growing public pressure to strengthen agri-environmental and environmental policies that can reduce the human health and environmental costs while increasing the environmental benefits associated with agriculture.
4. Agricultural policy reforms with a continued shift towards decoupled support and measures aimed at environmental improvement on-farms.
5. Innovations in policy and market approaches to address environmental issues in agriculture, that seek to change the behaviour of farmers, the agro-food chain and other stakeholders to improve environmental quality, for example, water treatment companies and/or community groups working with farmers to address agricultural water pollution (OECD, 2012c).

The environmental performance of OECD agriculture over the past decade examined in this report, provides some indication that agriculture and policy makers are capable of meeting the future economic, social and environmental challenges for the sector. Examples include efficiency and management improvements in the use of nutrients, pesticides and water resources, and enhancing environmental benefits that can stem from certain management practices, such as conservation tillage and riparian buffers along water courses. But there are signs in regions of some OECD countries where progress in improving environmental performance has been disappointing and more effort is required from all stakeholders, for example, with water pollution and the decline in farmland breeding bird populations.

References

European Environment Agency (2010), *The European Environment State and Outlook 2010: Freshwater Quality*, Copenhagen, Denmark, *www.eea.europea.eu*.

National Research Council (2008), "Mississippi River Water Quality and the Clean Water Act: Progress, Challenges and Opportunities", *The National Academies Press*, Washington DC, United States, *www.nap.edu/catalog/12051.html*.

OECD (2012a), *OECD-FAO Agricultural Outlook 2012-2021*, OECD Publishing, *www.agri-outlook.org*.

OECD (2012b), *Agricultural Policy Monitoring and Evaluation 2012*, OECD Publishing, *www.oecd.org/agr*.

OECD (2012c), *Water Quality and Agriculture: Meeting the Policy Challenge*, OECD Publishing, *www.oecd.org/agriculture/water*.

OECD (2010a), *Linkages Between Agricultural Policies and Environmental Effects: Using the OECD Stylised Agri-Environmental Policy Impact Model*, OECD Publishing.

OECD (2010b), *Agricultural Policies in OECD Countries: At a Glance*, OECD Publishing, *www.oecd.org/agr*.

PART II

OECD trends on environmental conditions related to agriculture since 1990

Chapter 3

Agricultural production, land use, organic farming and transgenic crops

This chapter reviews the environmental performances of agriculture in OECD countries related to agricultural production, land use, organic farming and transgenic crops. It provides a description of the policy context (issues and main challenges), definitions for the agri-environmental indicators presented, and elements related to concepts, interpretations, links to other indicators, as well as measurability and data quality. The chapter then describes the main trends of the agri-environmental indicators, using available data covering the period 1990-2010 and based on a set of tables and figures.

3.1. Policy context

The issue

Fundamental to the environmental performance of agriculture is the relationship between agricultural production and land use. Agricultural policies can influence production decisions by farmers, in terms of scale, intensity and the composition of production (Figure 1.1). This has changed significantly since 1990, across many OECD countries, as agricultural support has become increasingly decoupled from production and use of inputs (Chapter 2, Section 2.2, Figures 2.2 and 2.3).

The reform of agricultural policies for many OECD countries, and the consequences for agricultural production, have in turn had an influence on the overall area of land used in agriculture, as well as land cover (e.g. crops, pasture) for land remaining in agriculture. These changes have provided the incentive to remove production from the extensive margin over the past 20 years, especially the transfer of marginal farmland into forestry, but probably less so at the intensive margin where agricultural land is often under competition for other uses, especially urban settlement and for transport infrastructure.

The emergence of agri-environmental and environmental policies since the early 1990s has also, for many OECD countries, encouraged farmers to use agricultural land less intensively, for example, payments to remove land from production or to convert to organic management which may lower environmental pressure by eliminating chemical input use. Simultaneously a combination of market forces and the regulatory framework in some countries has facilitated an expansion of land cultivated to transgenic crops.

Main challenges

Agricultural land is not only a production factor for farming activities, but also an important source of social benefits and costs arising from the agricultural production, including positive externalities, such as landscape amenities, carbon storage, and the regulation of water flows; but also negative externalities, for example, chemical run-off or air emissions that impact on water and air quality.

Clearly the global challenge over the coming decades will be to raise agricultural production and productivity to meet the rise in world demand for food, feed, fibre, and renewable energy, while at the same time minimising the consequences for the environment and managing natural resources sustainably. This also needs to be achieved in the context of growing competition for land, water and other natural resources and increasing concerns for agriculture related to climate change, including climate variability.

3.2. Indicators

Definitions

The indicators related to agricultural production, land use, organic farming and transgenic crops are defined as changes in:

- Agricultural production volume: index of change in total agriculture, crop and livestock production.
- Agricultural land use area.
- Certified organic farming area.
- Transgenic crops area.

Concepts, interpretation, limitations and links to other indicators

The relationships between agricultural production, land use and environmental externalities are complex. This is because the modality and the extent to which agriculture produces environmental externalities, positive and negative, depends on multiple factors, as discussed in Chapter 2 and encapsulated in Figure 1.1. Hence, interpretation of the set of indicators in this chapter needs to be viewed against the overall environmental performance of agriculture.

While increases in agricultural production and land used for farming will in general tend to signify greater pressure on the environment, this will depend on the extent to which farming systems and practices can limit or overcome these pressures. Equally, where agricultural production increases but on a declining area of land, this may intensify production (crops/livestock) for a given unit of land which might also heighten environmental pressure depending on how production and the land is managed. As agriculture, however, is a major user of natural resources, it will inevitably lead to some disturbance of the environment as without the use of land, water and energy and other inputs, agriculture cannot produce food and other commodities.

Some caution is required in interpreting changes over time and between countries of indicators related to land under organic management and transgenic crops. The two key limitations to these indicators concern: first, definitional issues in terms of the consistency of what constitutes organic farming and a transgenic crop; and second, problems of clearly defining the environmental implications of an increase/decrease in the area under organic management or transgenic crops, as compared to other farm management systems, as examined later in this chapter.

The agricultural production and land indicators discussed in this chapter, provide the broad context to the environmental performance of agriculture, and are linked to all the other agri-environmental indicators examined in the report. The evolution of agricultural production and land use changes are key drivers on farm input use (e.g. nutrients, pesticides, energy and water resources) and their management. These drivers play a major role in affecting the state of the environment related to soils, water, air and biodiversity, which in turn impact on human welfare, including human health, social values, and agriculture itself, as well as other commercial activities, such as fishing (Figure 1.1).

Measurability and data quality

The FAO indices of agricultural production show the relative level of the aggregate volume of agricultural production for each year in comparison with the base period 2004-06.

They are based on the sum of price-weighted quantities of different agricultural commodities produced after deductions of quantities used as seed and feed weighted in a similar manner. The resulting aggregate represents, therefore, disposable production for any use except as seed and feed. All the indices at the country, regional and world levels are calculated by the Laspeyres formula. Production quantities of each commodity are weighted by 2004-06 average international commodity prices and summed for each year. To obtain the index, the aggregate for a given year is divided by the average aggregate for the base period 2004-06. Overall agricultural production data have a high degree of reliability as they are collected annually for most countries (Annex 1.A2).

Agricultural land use data, including land under organic management and transgenic crops, are also usually of high quality and reliability (Annex 1.A2). Definitional issues related to organic farming and transgenic crops, however, may be an issue in some cases (e.g. whether a farm is in the transition toward, or already fully certified as organic or where a farm is already organic but is not registered as such), as noted above.

3.3. Main trends

Growth in OECD ***agricultural production*** slowed over the decade from 2000 to 2010 compared to the 1990s (Figures 3.1 to 3.3). This development was in part explained by the slowdown in growth of production in many of the OECD agricultural exporting countries since 2000, notably **Canada**, **Chile**, **Mexico**, **New Zealand**, **Spain**, **Turkey** and the **United States**. **Israel** is an exception, with agricultural production volumes increasing at a more rapid annual rate over the 2000s compared to the 1990s. **Australia** experienced strong growth in agricultural production over the 1990s, but slowed and was highly variable since 2000, partly induced by major droughts (Figure 3.4). Agricultural production for most of these countries is projected by OECD-FAO to continue its upward trend over the next decade (Figure 2.6).

For most **European Union** countries growth in agricultural production slowed appreciably since 2000, and even decreased in some cases (e.g. **France**, **Greece**), compared to modest growth over the 1990s (Figure 3.1). The shock to the agricultural sector of EU countries in transition to a market economy over the 1990s led to a sharp contraction of production albeit with some recovery over the 2000s, especially for **Estonia**, **Hungary** (crops) and **Poland** (livestock) (Figures 3.2 and 3.3). Overall agricultural production for the **EU27** is projected to show a modest increase to 2020 (Figure 2.6).

The OECD average annual decrease of -0.3% per annum in the ***area of agricultural land*** over the 1990s, accelerated to a per annum reduction of -0.5% (Figure 3.5). But despite the overall trend to remove land from agricultural use it remains the major land use for many countries, with agriculture representing over 40% of the total land area for two-thirds of OECD countries by 2008-10 (Figure 3.6). The withdrawal of agricultural land from production has been most marked over the past twenty years for many of the EU transition countries (i.e. the **Czech and Slovak Republics**, **Estonia**, **Hungary**, **Poland** and **Slovenia**), but also notably in **Italy**, **Korea** and **Spain**.

Agricultural land area has expanded in only six OECD countries since over the 2000s (Figure 3.5). Where this has occurred it has mainly involved countries with rapidly expanding agricultural sectors (e.g. **Canada**, **Chile** and **Mexico** in the 1990s) or cases where better reporting of land farmed is partly due to improved registration systems linked to requirements for payments under some agri-environmental schemes and more generally

Figure 3.1. **Agricultural production volume index, OECD countries, 1990-2010**

Base 100 = 2004-06

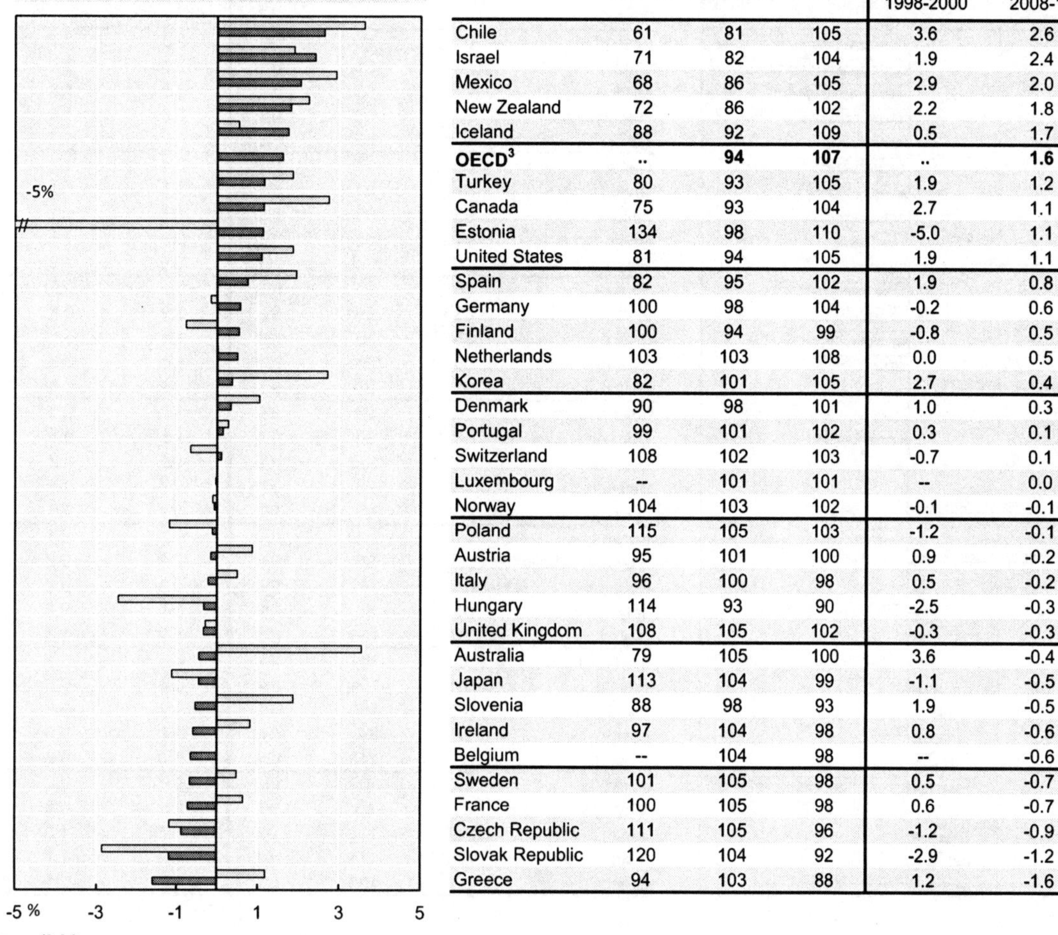

	Average			Average annual % change	
	1990-92[1]	1998-2000[2]	2008-10	1990-92 to 1998-2000	1998-2000 to 2008-10
Chile	61	81	105	3.6	2.6
Israel	71	82	104	1.9	2.4
Mexico	68	86	105	2.9	2.0
New Zealand	72	86	102	2.2	1.8
Iceland	88	92	109	0.5	1.7
OECD[3]	..	**94**	**107**	..	**1.6**
Turkey	80	93	105	1.9	1.2
Canada	75	93	104	2.7	1.1
Estonia	134	98	110	-5.0	1.1
United States	81	94	105	1.9	1.1
Spain	82	95	102	1.9	0.8
Germany	100	98	104	-0.2	0.6
Finland	100	94	99	-0.8	0.5
Netherlands	103	103	108	0.0	0.5
Korea	82	101	105	2.7	0.4
Denmark	90	98	101	1.0	0.3
Portugal	99	101	102	0.3	0.1
Switzerland	108	102	103	-0.7	0.1
Luxembourg	--	101	101	--	0.0
Norway	104	103	102	-0.1	-0.1
Poland	115	105	103	-1.2	-0.1
Austria	95	101	100	0.9	-0.2
Italy	96	100	98	0.5	-0.2
Hungary	114	93	90	-2.5	-0.3
United Kingdom	108	105	102	-0.3	-0.3
Australia	79	105	100	3.6	-0.4
Japan	113	104	99	-1.1	-0.5
Slovenia	88	98	93	1.9	-0.5
Ireland	97	104	98	0.8	-0.6
Belgium	--	104	98	--	-0.6
Sweden	101	105	98	0.5	-0.7
France	100	105	98	0.6	-0.7
Czech Republic	111	105	96	-1.2	-0.9
Slovak Republic	120	104	92	-2.9	-1.2
Greece	94	103	88	1.2	-1.6

--: not available.

Notes: Countries are ranked in terms of highest to lowest % annum growth rate 1998-2000 to 2008-10.

The FAO indices of agricultural production show the relative level of the aggregate volume of agricultural production for each year in comparison with the base period 2004-06. They are based on the sum of price weighted quantities of different agricultural commodities produced after deductions of quantities used as seed and feed weighted in a similar manner. The resulting aggregate represents, therefore, disposable production for any use except as seed and feed. All the indices at the country, regional and world levels are calculated by the Laspeyres formula. Production quantities of each commodity are weighted by 2004-06 average international commodity prices and summed for each year. To obtain the index, the aggregate for a given year is divided by the average aggregate for the base period 2004-06.

The statistical data for Israel are supplied by and under the responsibility of the relevant Israeli authorities. The use of such data by the OECD is without prejudice to the status of the Golan Heights, East Jerusalem and Israeli settlements in the West Bank under the terms of international law.

1. Data for 1990-92 average equal the 1993-95 average for the Czech Republic and the Slovak Republic; and the 1992-94 average for Estonia and Slovenia.
2. Data for 1998-2000 average equal the 2000-02 average for Belgium, Luxembourg and total OECD average.
3. The OECD aggregate includes Australia, Canada, Chile, EU27, Iceland, Israel, Japan, Korea, Mexico, New Zealand, Norway, Switzerland, Turkey and United States.

Source: FAOSTAT (2012), *http://faostat.fao.org*.

StatLink http://dx.doi.org/10.1787/888932792483

agricultural support policies, especially in EU countries (e.g. **Finland**, **Greece**, **Luxembourg**, **Sweden** and over the 1990s **Belgium**, **Denmark** and **Norway**).

The relationship between changes in agricultural production and agricultural land can provide a broad indication of the trends in the environmental performance of agriculture as a backdrop to the discussion of more detailed indicators in following chapters of the report (Figure 3.7). Countries can be grouped under four main categories according to

Figure 3.2. **Crop production volume index, OECD countries, 1990-2010**

Base 100 = 2004-06

	Average			Average annual % change	
	1990-92[1]	1998-2000[2]	2008-10	1990-92 to 1998-2000	1998-2000 to 2008-10
Iceland	62	77	101	2.8	2.7
Chile	64	81	103	3.0	2.4
Mexico	73	89	104	2.5	1.6
New Zealand	75	93	109	2.8	1.6
OECD[3]	..	**95**	**106**	..	**1.2**
Canada	80	97	109	2.4	1.2
United States	81	93	104	1.7	1.1
Finland	93	89	100	-0.6	1.1
Netherlands	91	95	105	0.5	1.1
Estonia	132	106	117	-3.6	1.0
Turkey	80	94	102	2.1	0.9
Israel	83	89	97	0.9	0.8
Spain	85	96	104	1.4	0.8
Austria	86	96	102	1.4	0.6
Hungary	98	85	90	-1.8	0.6
Luxembourg	--	96	98	--	0.3
Germany	87	98	101	1.5	0.3
Denmark	105	103	106	-0.3	0.2
Belgium	--	96	97	--	0.1
Norway	113	95	95	-2.1	0.0
Korea	91	100	100	1.2	-0.1
Slovak Republic	111	96	94	-2.9	-0.2
Portugal	108	101	99	-0.8	-0.2
Italy	95	98	95	0.4	-0.2
United Kingdom	107	106	103	-0.2	-0.3
Czech Republic	100	101	96	0.2	-0.5
France	97	104	99	0.8	-0.5
Sweden	105	105	100	0.0	-0.6
Ireland	94	97	91	0.4	-0.6
Poland	122	112	104	-1.0	-0.8
Australia	69	109	100	5.9	-0.8
Slovenia	85	98	90	2.5	-0.9
Japan	120	106	96	-1.5	-1.0
Switzerland	118	113	101	-0.5	-1.2
Greece	92	105	85	1.7	-2.1

--: not available.

Notes: Countries are ranked in terms of highest to lowest % annum growth rate 1998-2000 to 2008-10.

The FAO indices of agricultural crop production show the relative level of the aggregate volume of agricultural crop production for each year in comparison with the base period 2004-06. They are based on the sum of price weighted quantities of different agricultural commodities produced after deductions of quantities used as seed and feed weighted in a similar manner. The resulting aggregate represents, therefore, disposable production for any use except as seed and feed. All the indices at the country, regional and world levels are calculated by the Laspeyres formula. Production quantities of each commodity are weighted by 2004-06 average international commodity prices and summed for each year. To obtain the index, the aggregate for a given year is divided by the average aggregate for the base period 2004-06.

The statistical data for Israel are supplied by and under the responsibility of the relevant Israeli authorities. The use of such data by the OECD is without prejudice to the status of the Golan Heights, East Jerusalem and Israeli settlements in the West Bank under the terms of international law.

1. Data for 1990-92 average equal the 1993-95 average for the Czech Republic and the Slovak Republic; and the 1992-94 average for Estonia and Slovenia.
2. Data for 1998-2000 average equal the 2000-02 average for Belgium and Luxembourg.
3. The OECD aggregate includes Australia, Canada, Chile, EU27, Iceland, Israel, Japan, Korea, Mexico, New Zealand, Norway, Switzerland, Turkey and United States.

Source: FAOSTAT (2012), *http://faostat.fao.org*.

StatLink http://dx.doi.org/10.1787/888932792502

production and land trends with varying implications for the environment between 1998-00 and 2008-10: (only a few countries are highlighted under each group as those most illustrative of the group shown in Figure 3.7):

- Group I – *Increasing production and expanding land area*: The risk of environmental pressure has likely been increasing for this group of countries, especially countries such as **Canada** and **Chile**. However, the overall balance of the environmental performance for these

Figure 3.3. **Livestock production volume index, OECD countries, 1990-2010**

Base 100 = 2004-06

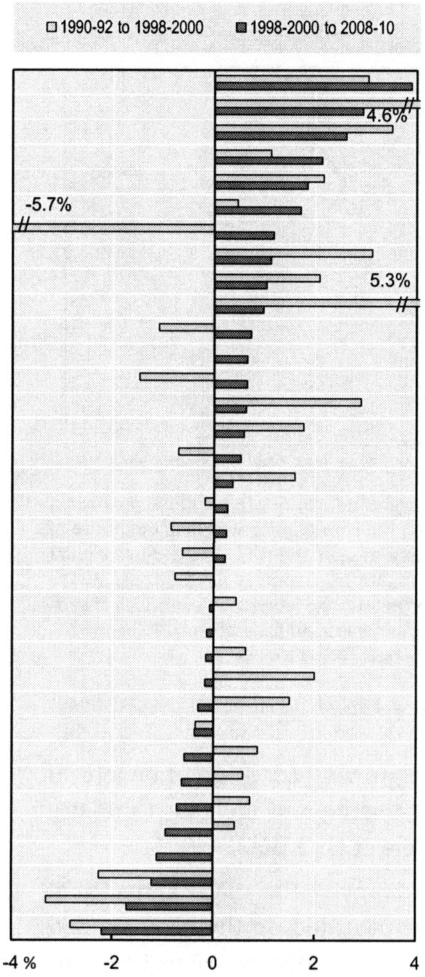

	Average			Average annual % change	
	1990-92[1]	1998-2000[2]	2008-10	1990-92 to 1998-2000	1998-2000 to 2008-10
Israel	60	76	111	3.0	3.9
Chile	57	82	109	4.6	2.9
Mexico	63	82	107	3.5	2.6
Turkey	84	91	113	1.1	2.1
New Zealand	71	85	102	2.2	1.8
Iceland	89	93	110	0.5	1.7
Estonia	134	94	106	-5.7	1.2
Canada	69	88	98	3.1	1.1
United States	81	95	106	2.1	1.0
Korea	68	103	113	5.3	1.0
Germany	108	99	106	-1.1	0.7
OECD[3]	..	**98**	**103**	..	**0.7**
Poland	108	96	103	-1.5	0.7
Spain	75	94	100	2.9	0.6
Portugal	87	101	107	1.8	0.6
Switzerland	105	99	104	-0.7	0.5
Denmark	84	96	99	1.6	0.4
Netherlands	108	107	110	-0.2	0.3
Finland	103	96	98	-0.9	0.2
Greece	99	95	97	-0.6	0.2
Japan	108	101	101	-0.8	0.0
Norway	101	105	104	0.4	-0.1
Luxembourg	--	102	101	--	-0.1
Italy	99	104	103	0.6	-0.2
Australia	87	102	100	2.0	-0.2
Slovenia	89	98	95	1.5	-0.3
United Kingdom	108	105	101	-0.4	-0.4
Ireland	98	105	99	0.9	-0.6
Austria	100	105	99	0.6	-0.6
Sweden	98	104	97	0.7	-0.7
France	103	106	96	0.4	-0.9
Belgium	--	109	100	--	-1.1
Czech Republic	121	108	95	-2.3	-1.2
Hungary	142	109	91	-3.3	-1.7
Slovak Republic	129	112	89	-2.8	-2.2

--: not available.

Notes: Countries are ranked in terms of highest to lowest % annum growth rate 1998-2000 to 2008-10.

The FAO indices of agricultural production show the relative level of the aggregate volume of agricultural production for each year in comparison with the base period 2004-06. They are based on the sum of price weighted quantities of different agricultural commodities produced after deductions of quantities used as seed and feed weighted in a similar manner. The resulting aggregate represents, therefore, disposable production for any use except as seed and feed. All the indices at the country, regional and world levels are calculated by the Laspeyres formula. Production quantities of each commodity are weighted by 2004-06 average international commodity prices and summed for each year. To obtain the index, the aggregate for a given year is divided by the average aggregate for the base period 2004-06.

The statistical data for Israel are supplied by and under the responsibility of the relevant Israeli authorities. The use of such data by the OECD is without prejudice to the status of the Golan Heights, East Jerusalem and Israeli settlements in the West Bank under the terms of international law.

1. Data for 1990-92 average equal the 1993-95 average for the Czech Republic and the Slovak Republic; and the 1992-94 average for Estonia and Slovenia.
2. Data for 1998-2000 average equal the 2000-02 average for Belgium, Luxembourg and OECD average.
3. The OECD aggregate for crop production indices included Australia, Canada, Chile, EU27, Iceland, Israel, Japan, Korea, Mexico, New Zealand, Norway, Switzerland, Turkey and United States.

Source: FAOSTAT (2012) *http://faostat.fao.org*.

StatLink *http://dx.doi.org/10.1787/888932792521*

countries will have depended on, for example, the change in quality of biodiversity of the land that has been brought into production (usually forested land), and also the extent of adoption of farm management practices that are beneficial to the environment. In **Canada**, for example, while agricultural nitrogen surpluses (Figures 4.1 and 4.2) and ammonia emissions (Figure 10.2; Table 10.1) have been increasing with potential threat to

Figure 3.4. **Agricultural production volume index, Australia, Canada and Spain, 1990-2010**

Base 100 = 2004-06

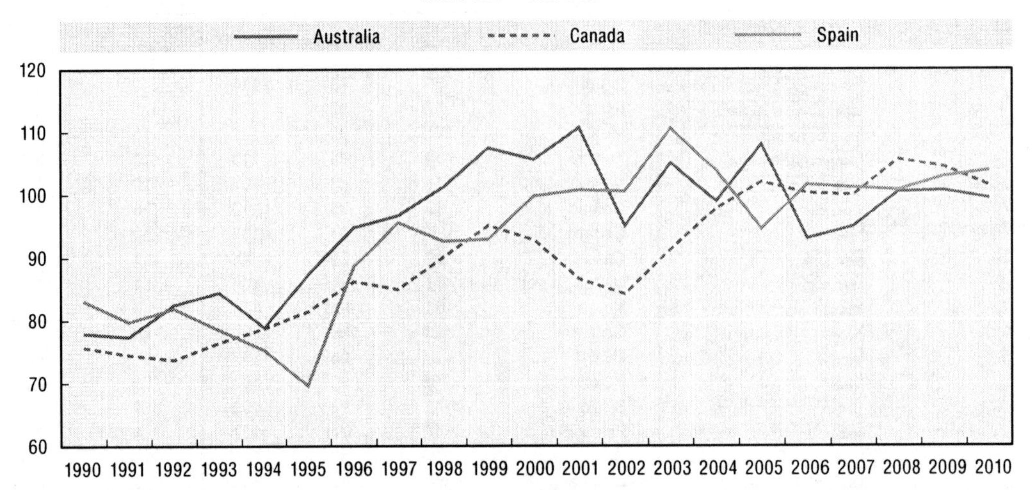

Notes: The FAO indices of agricultural production show the relative level of the aggregate volume of agricultural production for each year in comparison with the base period 2004-06. They are based on the sum of price weighted quantities of different agricultural commodities produced after deductions of quantities used as seed and feed weighted in a similar manner. The resulting aggregate represents, therefore, disposable production for any use except as seed and feed. All the indices at the country, regional and world levels are calculated by the Laspeyres formula. Production quantities of each commodity are weighted by 2004-06 average international commodity prices and summed for each year. To obtain the index, the aggregate for a given year is divided by the average aggregate for the base period 2004-06.

Source: FAOSTAT (2012), *http://faostat.fao.org*.

StatLink http://dx.doi.org/10.1787/888932792540

water and air quality, at the same time there has been growing adoption of beneficial nutrient management practices, such as soil nutrient testing and increased manure storage capacity and adoption of conservation tillage (Eilers et al., 2010).

- Group II – *Increasing production on a decreasing land area:* Many of the OECD agricultural exporting countries fall within this group, as over the past decade they have raised agricultural productivity on a reduced land area (e.g. **Israel**, **Mexico**, **New Zealand**, **Spain**, **Turkey** and the **United States**). It is difficult to generalise as to the overall environmental performance of agriculture in these countries. Agricultural production increases on a declining area of land may intensify production (crops/livestock) for a given unit of land and heighten environmental pressure. This will depend, however, on how production is managed, for while these countries have been increasing their input use (fertilisers, pesticides, water resource use and energy) and livestock density rates tending to exert greater environmental pressure, this is also being offset to some extent by adoption of farm management practices beneficial to the environment. Water application rates per hectare irrigated, for example, have improved in **Israel**, **Mexico**, **Spain** and the **United States** (Figure 8.2).
- Group III – *Decreasing production and land area*: Most of the countries in this group are in the **European Union**, but also includes **Japan**, where the overall decrease in growth rate of agricultural production on a reduced land area has tended to lower environmental pressure. Nevertheless, for some agri-environmental indicators these countries remain above the OECD average (e.g. nutrient surpluses per hectare, Figures 4.2 and 4.4), and continue to exert considerable pressure on environmental quality, even if at diminishing levels over the past decade. **Australia**, stands out in this group as production volumes have been highly

Figure 3.5. **Agricultural land area, OECD countries, 1990-2010**

	Average (thousand hectares)			Average annual % change	
	1990-92[1]	1998-2000[2]	2008-10[3]	1990-92 to 1998-2000	1998-2000 to 2008-10
Greece	3 661	3 583	4 076	-0.2	1.9
Chile	15 748	15 130	15 738	-0.5	0.4
Finland	2 542	2 204	2 295	-1.8	0.4
Luxembourg	126	127	131	0.2	0.3
Estonia	1 369	911	929	-5.0	0.2
Canada	62 215	62 552	63 427	0.1	0.2
Sweden	3 370	3 041	3 075	-1.3	0.1
Turkey	40 673	39 094	39 015	-0.5	0.0
United States	426 442	414 696	413 693	-0.3	0.0
Iceland	2 416	2 409	2 403	0.0	0.0
United Kingdom	18 249	17 717	17 421	-0.4	-0.2
Switzerland[4]	1 067	1 074	1 055	0.1	-0.2
France	30 365	29 805	29 272	-0.2	-0.2
Belgium	1 371	1 393	1 366	0.2	-0.2
Germany	17 373	17 197	16 840	-0.1	-0.2
Ireland	4 464	4 425	4 320	-0.1	-0.2
Netherlands	1 985	1 959	1 909	-0.2	-0.3
Norway	1 002	1 042	1 015	0.5	-0.3
Slovenia	557	500	481	-1.5	-0.4
Mexico	104 500	106 300	102 705	0.2	-0.4
EU15	**141 648**	**134 135**	**128 376**	**-0.7**	**-0.4**
OECD	**1 314 767**	**1 286 623**	**1 222 442**	**-0.3**	**-0.5**
Portugal	4 024	3 885	3 684	-0.4	-0.5
Japan	5 204	4 867	4 610	-0.8	-0.5
Hungary	6 356	6 078	5 703	-0.6	-0.6
Austria	3 468	3 380	3 193	-0.3	-0.7
Denmark	2 776	2 874	2 672	0.4	-0.8
Korea	2 179	1 953	1 807	-1.4	-0.9
Israel	578	562	519	-0.3	-0.9
Spain	30 226	26 870	24 209	-1.5	-1.0
Australia	464 367	457 677	408 299	-0.2	-1.1
New Zealand	13 151	12 710	11 264	-0.4	-1.2
Poland	18 594	18 224	15 926	-0.3	-1.3
Italy	17 647	15 673	13 914	-1.5	-1.5
Czech Republic	4 285	4 279	3 547	0.0	-1.9
Slovak Republic	2 417	2 430	1 929	0.1	-2.3

Notes: Countries are ranked from highest to lowest % annum growth rate 1998-2000 to 2008-10.

Agricultural land is defined as arable and permanent cropland plus permanent and temporary pasture.

The statistical data for Israel are supplied by and under the responsibility of the relevant Israeli authorities. The use of such data by the OECD is without prejudice to the status of the Golan Heights, East Jerusalem and Israeli settlements in the West Bank under the terms of international law.

1. Data for 1990-92 average equal the 1991-93 average for Slovenia; and the year 1990 for Greece and Switzerland.
2. Data for 1998-2000 average equal the 1999-01 average for Austria; and the year 2000 for Greece.
3. Data for 2008-10 average equal the 2007-09 average for Austria, Canada, Chile, Denmark, Iceland, Israel, Italy, Korea; Mexico; and the year 2007 for Greece.
4. In the case of Switzerland, data refer to Utilised Agricultural Area (hectares), including arable and permanent cropland, but excluding summer pasture.

Source: FAOSTAT (2012), *http://faostat.fao.org and national data.*

StatLink http://dx.doi.org/10.1787/888932792559

variable since 2000, partly induced by drought and volatile world commodity market prices. According to a recent study the situation and outlook for the environmental performance of Australian agriculture is mixed, with progress in many aspects of land management (e.g. by 2010, the extent of land clearing was balanced by the extent of regrowth) but in other areas environmental indicators remain adverse, such as acidification and wind erosion of agricultural soils (State of the Environment Committee, 2011).

- Group IV – *Decreasing production but on expanding land area*: Only **Greece**, **Luxembourg** and **Sweden** are in this group, with over the past decade agricultural production declining but on an expanding land area, which would suggest there has been a trend toward the extensification of agriculture. It is likely, however, that the expansion in the agricultural

Figure 3.6. **Agricultural land use in the national land area, OECD countries, 2008-10**

% share average 2008-10

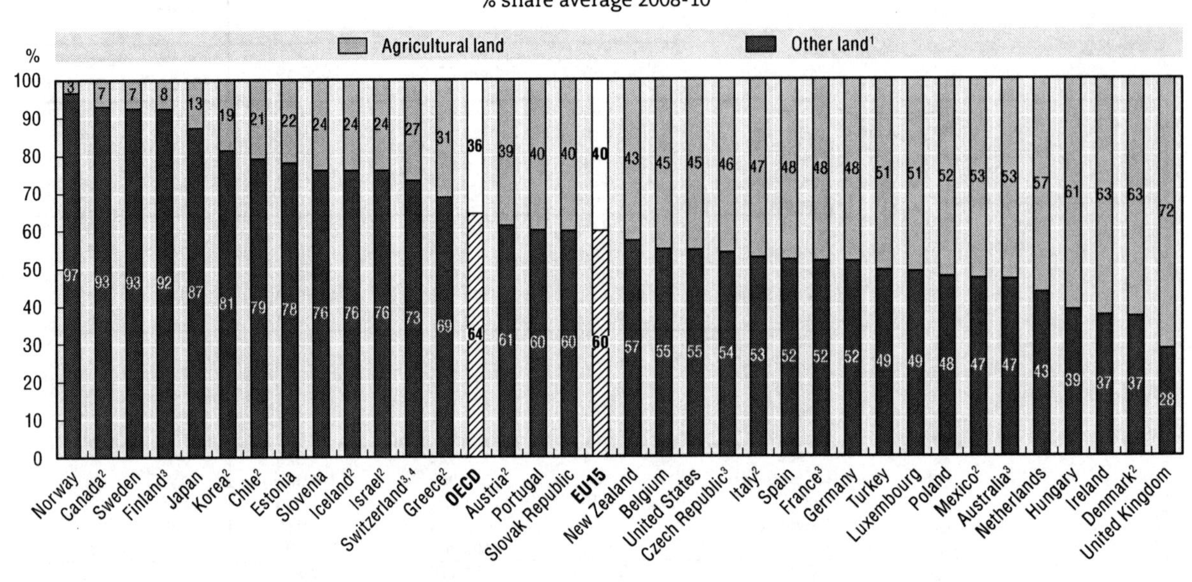

Notes: Countries are ranked from the lowest to the highest share of agricultural land in national land area.

National land area excluding water bodies (e.g. rivers, lakes, etc.).

The statistical data for Israel are supplied by and under the responsibility of the relevant Israeli authorities. The use of such data by the OECD is without prejudice to the status of the Golan Heights, East Jerusalem and Israeli settlements in the West Bank under the terms of international law.

1. Other land includes mainly forestry and urban land.
2. Data for 2008-10 agricultural land area average equal to the 2007-09 average for Austria, Canada, Chile, Denmark, Iceland, Israel, Korea and Mexico; the 2007-08 average for Italy; and the year 2007 for Greece.
3. Data for national land area 2008-10 average equal to the 2008-09 average for Australia, Czech Republic, Finland, France and Switzerland; and the year 2007 for Greece.
4. In the case of Switzerland, data refer to Utilised Agricultural Area (hectares), including arable and permanent cropland, but excluding summer pastures.

Source: FAOSTAT (2012), *http://faostat.fao.org and national data.*

StatLink http://dx.doi.org/10.1787/888932792578

land area for both these countries has been mainly the result of better reporting of land farmed, partly due to improved registration systems linked to requirements for payments under some EU agri-environmental and agricultural policy schemes. This was also the case for **Norway** over the 1990s (Figure 3.7; and OECD, 2008).

The share of agricultural land under certified organic farming remains very low across OECD countries below 2% for the OECD average (2008-10) (Figure 3.8). But this masks substantial variation across countries with shares tending to be higher than the OECD average in mainly **EU** countries, and below the average for most non-EU countries (Figure 3.8). To some extent this reflects varying policy environments, for example, with organic conversion payments provided to EU farmers, but not available to farmers in countries such as **Australia**, **Canada**, **Chile**, **Israel** and **New Zealand** (Vojtěch, 2010). This is also reflected in the variable growth in organic farming from 2002 to 2010, with growth more rapid in mainly **European** OECD countries (e.g. **Austria**, **Czech and Slovak Republics**, **Estonia** and **Sweden**), and less rapid in largely non-European OECD countries (e.g. **Japan**, **Mexico**).

Organic farming systems usually involve practices that maintain or improve the physical, chemical and biological conditions of soil, compared to other farming systems (OECD, 2012). Organic farming practices can also bring other benefits, such as to water quality by not using synthetic pesticides, as well as providing other ecosystem services, for example, carbon sequestration and enhanced biodiversity (Greene et al., 2009; OECD, 2003; and Stolze et al., 2000).

Figure 3.7. **Agricultural production volume index and agricultural land area, OECD countries, 1990-2010**

Average annual % change 1990-1992[3] to 1998-2000[4]

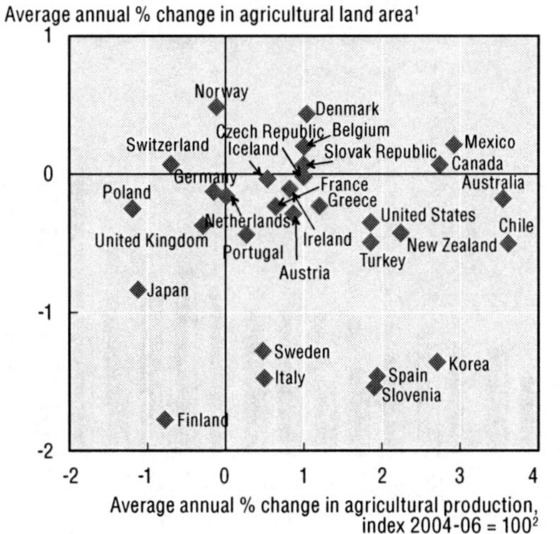

Average annual % change 1998-2000[4] to 2008-2010[5]

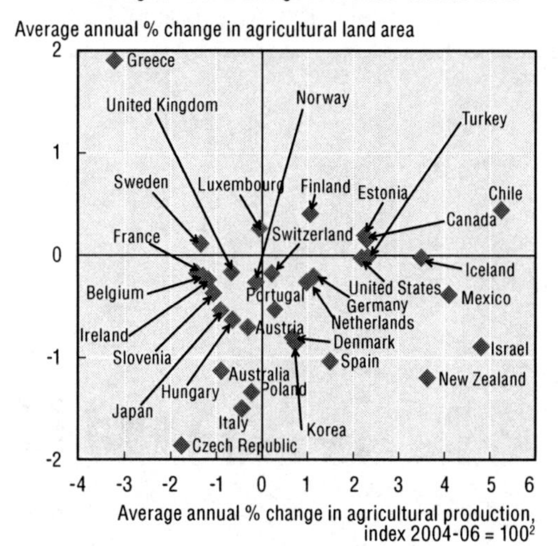

Note: The statistical data for Israel are supplied by and under the responsibility of the relevant Israeli authorities. The use of such data by the OECD is without prejudice to the status of the Golan Heights, East Jerusalem and Israeli settlements in the West Bank under the terms of international law.

1. Agricultural land is defined as arable and permanent cropland plus permanent and temporary pasture.
2. The FAO indices of agricultural production show the relative level of the aggregate volume of agricultural production for each year in comparison with the base period 2004-06. They are based on the sum of price weighted quantities of different agricultural commodities produced after deductions of quantities used as seed and feed weighted in a similar manner. The resulting aggregate represents, therefore, disposable production for any use except as seed and feed. All the indices at the country, regional and world levels are calculated by the Laspeyres formula. Production quantities of each commodity are weighted by 2004-06 average international commodity prices and summed for each year. To obtain the index, the aggregate for a given year is divided by the average aggregate for the base period 2004-06.
3. Agricultural land area data for 1990-92 average equal to the year 1990 for Greece and Switzerland; the year 1992 for Estonia; and the 1992-93 average for Slovenia. Data for agricultural production index 1990-92 average are not available for Belgium, Czech Republic, Luxembourg and Slovak Republic.
4. Agricultural land area data for 1998-2000 average equal to the 1999-2001 average for Austria; and to the year 2000 for Belgium, Greece and Luxembourg.
5. Agricultural land area data for 2008-10 average equal to the 2007-09 average for Austria, Canada, Chile, Denmark, Iceland, Korea and Mexico; the year 2007 for Greece; and the 2007-08 average for Italy. The Slovak Republic is not included in the figure, with average annual % changes 1998-2000 to 2008-10 in, respectively, agricultural production index and agricultural land area, equal to 0.1% and -2.3%.

Source: FAOSTAT (2012), *http://faostat.fao.org and national data.*

StatLink http://dx.doi.org/10.1787/888932792597

Most organic systems limit or prohibit all forms of chemical input use, so from this viewpoint are likely to lessen environmental pressure. But there are situations where intensive management within organic farming can lead to livestock manure, for example, being applied in excess of requirements (OECD, 2003). Farm comparisons with conventional farming show that actual nitrogen leaching rates per hectare can be up to almost 60% lower on organic than on conventional fields, but leaching rates per kilogram of output were similar or slightly higher.

Critical areas that can lead to higher nitrate leaching in organic farming are, for example: ploughing legumes at the wrong time; the selection of unfavourable crops planted in a rotation; and composting farmyard manure on unpaved surfaces (Stolze et al., 2000). Organic farming also often involves increased tillage to manage weeds (in the absence of pesticides), which may increase soil erosion.

Figure 3.8. **Agricultural land area under certified organic farm management, OECD countries, 2002-10**

% share of agricultural land area

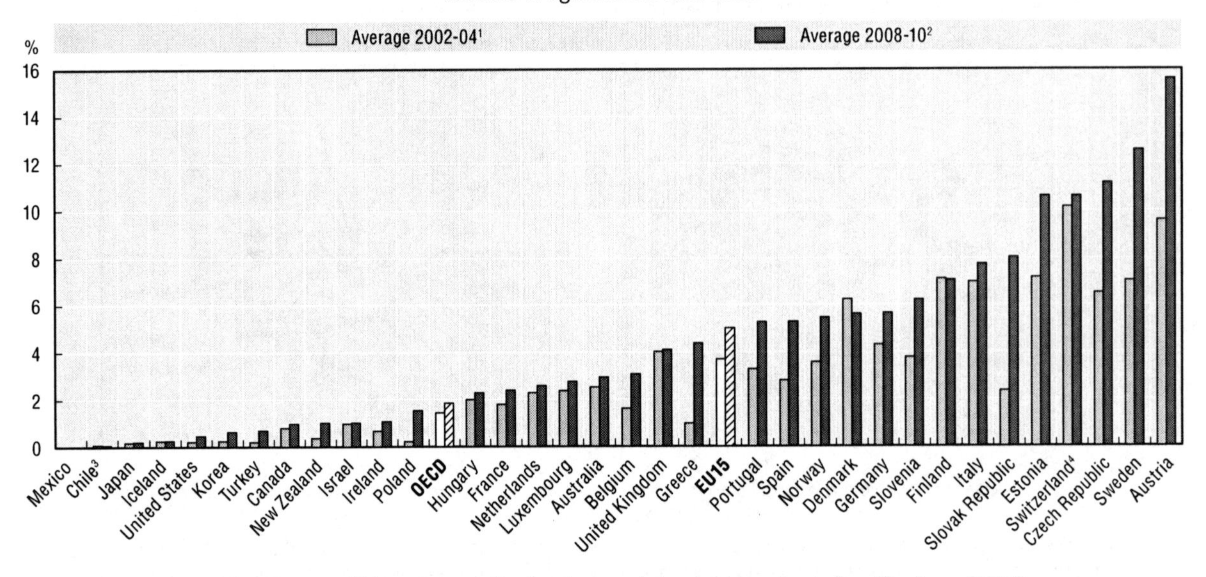

Notes: Countries are ranked from lowest to highest % share of land under organic farming in total agricultural land area 2008-10.

The statistical data for Israel are supplied by and under the responsibility of the relevant Israeli authorities. The use of such data by the OECD is without prejudice to the status of the Golan Heights, East Jerusalem and Israeli settlements in the West Bank under the terms of international law.

1. Data for 2002-04 average equal to the year 2005 for Estonia and Japan; the 2003-04 average for Chile and Korea; the 2003-05 average for Israel and Poland; and the year 2003 for Greece.
2. Data for the 2008-10 average equal to the 2007-09 average for Austria, Canada, Chile, Denmark, Iceland, Israel, Korea, Mexico and Spain; the 2007-08 average for Italy; and the year 2007 for Greece.
3. Data for Chile exclude wild harvesting areas and forests.
4. In the case of Switzerland, organic farming as a share of the Utilised Agriculture Area (hectares), including arable and permanent cropland, but excluding summer pastures.

Source: IFOAM (International Federation of Organic Agriculture Movements), *www.organic-world.net/statistics-data-sources.html*; Statistical Office of the European Community (EUROSTAT, see *http://epp.eurostat.ec.europa.eu*); and national data.

StatLink http://dx.doi.org/10.1787/888932792616

Some 18% of the total OECD arable and permanent cropland area is sown to transgenic crops in 2008-10 (sometimes referred to as *genetically modified crops*) (Table 3.1). The **United States** dominates OECD commercial production of transgenic crops. Regulations in **European** OECD countries and **Korea**, prevent the commercial exploitation of these crops, with only small areas sown for experimental purposes.

The OECD area of transgenic crops has grown rapidly since the mid-1990s, especially in **Canada** and the **United States**, dominated by herbicide tolerant crops (soybean, maize, canola, and cotton). OECD countries account for slightly more than half of the world global planted area of transgenic crops, but countries such as Argentina, Brazil, China and India have expanded use of these crops substantially over the past decade (Table 3.1).

The development of transgenic crops has led to ongoing discussions and debate on the potential environmental costs and benefits of using these crops as well as safety for human health. For example concerns have been raised over the possibility of genetic mingling of traditional species and wild relatives, such as maize in **Mexico** (OECD, 2005). **Mexico** is recognised as a "Vavilov" centre, which is an area where crops were first domesticated and have evolved over several thousand years, as is the case for maize (OECD, 2008). At the same time some researchers view transgenic crops as bringing benefits in terms, for example, of reducing pesticide use or providing crops with water saving traits.

Table 3.1. Transgenic crops, OECD and other major producing countries, 1996-2011

Area thousand hectares

	1996	1997	1998	1999	2000	2001	2002	2003	2004	2005	2006	2007	2008	2009	2010	2011	% share in total arable and permanent crops area (average 2008-10[1])
OECD member countries																	
United States	1 500	8 100	20 500	28 700	30 300	35 700	39 000	42 800	47 600	49 800	54 600	57 700	62 500	64 000	66 800	69 000	39.0
Canada	100	1 300	2 800	4 000	3 000	3 200	3 500	4 400	5 400	5 800	6 100	7 000	7 600	8 200	8 900	10 400	19.5
Australia	< 50	100	100	100	200	200	100	100	200	300	200	100	200	200	700	700	1.4
Chile[2]	..	..	..	..	..	..	11	9	9	13	19	24	30	25	20	< 100	1.4
Spain	..	..	< 50	< 50	< 50	< 50	< 50	< 50	100	100	100	100	100	100	100	100	0.6
Mexico	< 50	< 50		< 50	< 50	< 50	< 50	< 50	100	100	100	100	100	100	100	200	0.4
Portugal	..	..	..	< 50	..	..	..	..	..	1	1	4	5	5	5	< 1	0.3
Germany	..	..	..	..	..	..	..	< 1	< 1	0	1	3	3	< 1	< 1	< 1	0.0
OECD total	**1 600**	**9 500**	**23 400**	**32 800**	**33 500**	**39 100**	**42 611**	**47 309**	**53 409**	**56 114**	**61 121**	**65 031**	**70 538**	**72 630**	**76 625**	**80 400**	**18.4**
OECD non member countries																	
Brazil	..	..	..	..	..	..	..	3 000	5 000	9 400	11 500	15 000	15 800	21 400	25 400	30 300	30.4
Argentina	100	1 400	4 300	6 700	10 000	11 800	13 500	13 900	16 200	17 100	18 000	19 100	21 000	21 300	22 900	23 700	65.9
India	..	..	..	..	..	..	< 100	100	500	1 300	3 800	6 200	7 600	8 400	9 400	10 600	5.0
China	..	0	< 100	300	500	1 500	2 100	2 800	3 700	3 300	3 500	3 800	3 800	3 700	3 500	3 900	3.0
Paraguay	..	..	..	..	..	..	..	..	1 200	1 800	2 000	2 600	2 700	2 200	2 600	2 800	65.3
South Africa	..	..	< 100	100	200	200	300	400	500	500	1 400	1 800	1 800	2 100	2 200	2 300	13.2
Other countries[3]	..	..	..	< 100	< 100	< 100	< 100	< 100	400	500	700	800	1 800	2 300	5 400	6 000	..
Non member total	100	1 400	4 300	7 100	10 700	13 500	15 900	20 200	27 500	33 900	40 900	49 300	54 500	61 400	71 400	79 600	..
World total	**1 700**	**10 900**	**27 800**	**39 900**	**44 200**	**52 600**	**58 500**	**67 500**	**80 900**	**90 014**	**102 021**	**114 331**	**125 038**	**134 030**	**148 025**	**160 000**	..

..: not available.

Note: Countries are ranked from highest to lowest area of transgenic crops in 2011.

1. Data for 2008-10 average equal to the 2007-09 average for Argentina, Brazil, Canada, China, India, Mexico, Paraguay and South Africa; and the 2006-08 average for Chile.
2. The area corresponds to multiplication of seeds which must be exported.
3. Other countries include Bulgaria, Colombia, Honduras, Indonesia, Philippines, Romania, Ukraine and Uruguay.

Source: ISAAA (International Service for the Acquisition of Agri-biotech Applications), Ithaca, New York State, United States, *www.isaaa.org*; Statistical Office of the European Community (EUROSTAT) and FAOSTAT (2012), *http://faostat.fao.org*.

StatLink http://dx.doi.org/10.1787/888932793433

References

Eilers, W., R. MacKay, L. Graham and A. Lefebvre (eds.) (2010), "Environmental Sustainability of Canadian Agriculture", *Agri-Environmental Indicator Report Series*, Report #3, Agriculture and Agri-Food Canada, Ottawa, Canada, *http://publications.gc.ca/collections/collection_2011/agr/A22-201-2010-eng.pdf.*

Greene, C., C. Dimitri, B.-H. Lin, W.D. McBride, L. Oberholtzer and T. Smith (2009), "Emerging Issues in the U.S. Organic Industry", *Economic Information Bulletin Report*, Number 55, Economic Research Service, US Department of Agriculture, March, Washington, DC, United States.

OECD (2012), *Water Quality and Agriculture: Meeting the Policy Challenge*, OECD Publishing, *www.oecd.org/agriculture/water.*

OECD (2008), *Environmental Performance of Agriculture at a Glance*, OECD Publishing, *www.oecd.org/agriculture/env/indicators.*

OECD (2005), *Agriculture, Trade and the Environment: Arable Crops Sector*, OECD Publishing.

OECD (2003), *Organic Agriculture: Sustainability, Markets and Policies*, OECD Publishing, *www.oecd.org/tad/env.*

State of the Environment 2011 Committee (2011), *Australia State of the Environment 2011*, Department of Sustainability, Environment, Water, Population and Communities, Australian Government, Canberra, Australia.

Stolze, M., A. Piorr, A. Häring and S. Dabbert (2000), *The Environmental Impacts of Organic Farming in Europe*, University of Hohenheim, Stuggart, Germany, *www.uni-hohenheim.de/i410a/ofeurope/organicfarmingineurope-vol6.pdf.*

Chapter 4

Nutrients: Nitrogen and phosphorus balances

This chapter reviews the environmental performances of agriculture in OECD countries related to nitrogen and phosphorus balances. It provides a description of the policy context (issues and main challenges), definitions for the agri-environmental indicators presented, and elements related to concepts, interpretations, links to other indicators, as well as measurability and data quality. The chapter then describes the main trends of the agri-environmental indicators, using available data covering the period 1990-2010 and based on a set of tables and figures.

4.1. Policy context

The issue

Inputs of nutrients, such as nitrogen and phosphorus, are necessary in farming systems as they are critical in maintaining and raising crop and forage productivity. Where nutrients are in deficit soil fertility can decline, while with an excess of nutrients necessary for plant growth there is a risk of polluting soil, air, and water (eutrophication). OECD agriculture is a significant source of nitrogen and phosphorus entering the environment as there is in most cases a surplus of nutrients compared to plant requirements. This concerns nearly all OECD countries, to varying degrees, and as a result there is an extensive range of policy instruments (payments, taxes, regulations, farm advice, etc.) used by countries to address nutrient pollution of water (Chapter 9) and air in terms of ammonia emissions (Chapter 10) (OECD, 2012).

Main challenges

A build up of surplus nutrients in excess of immediate crop and forage needs can lead to nutrient losses representing not only a possible cause of economic inefficiency in nutrient use by farmers, but especially a source of potential harm to the environment. This can occur in terms of water pollution (e.g. eutrophication of surface water caused by nutrient runoff and groundwater pollution by leaching), and air pollution, notably ammonia, as well as greenhouse gas emissions. An additional environmental issue concerns the sustainability of phosphorus resources, as world reserves are finite, although this could induce better management and recycling of phosphorus in agriculture.

Agricultural activities will usually involve some loss of nutrients into the environment, as it is technically impossible to achieve zero pollution in most situations. Even in pristine water environments natural sources (e.g. soil minerals) can cause changes in the physical, chemical and biological characteristics of water. The challenge in agriculture is to seek ways to increase production while minimising farm nutrient losses and subsequent damage to the environment (OECD, 2012).

4.2. Indicators

Definitions

The indicator related to agricultural nutrient balances include changes in:

- Gross agricultural nitrogen (N) and phosphorus (P) balances, surplus or deficit.

Concepts, interpretation, limitations and links to other indicators

The gross nutrient balances (N and P) are calculated as the difference between the total quantity of nutrient inputs entering an agricultural system (mainly fertilisers, livestock manure), and the quantity of nutrient outputs leaving the system (mainly uptake of nutrients by crops and grassland), as elaborated in Figure 4.1 (OECD/EUROSTAT, 2012a; 2012b). This calculation can be used as a proxy to reveal the status of environmental

Figure 4.1. **Main elements in the gross nitrogen and phosphorus balance calculation**

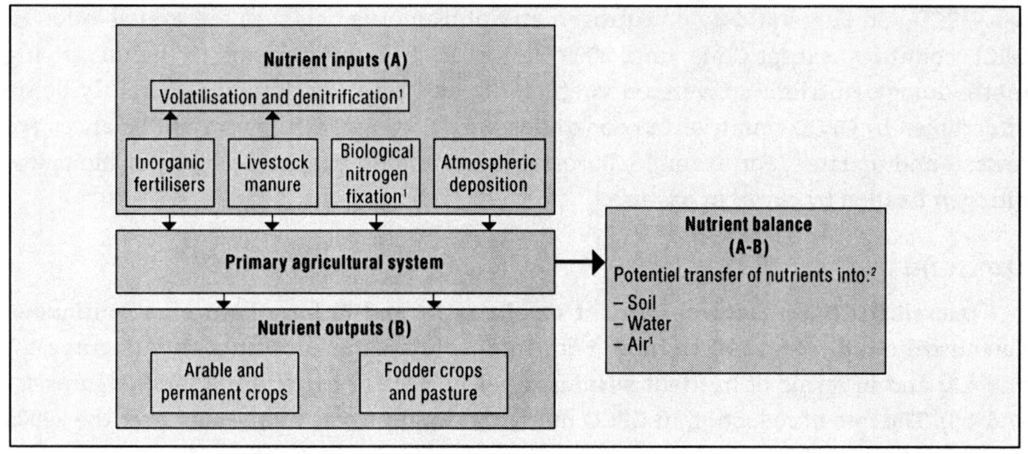

1. Applies to the nitrogen balance only.
2. Nutrients surplus to crop/pasture requirements are transported into the environment, potentially polluting soils, water and air, but a deficit of nutrients in soils can also occur to the detriment of soil fertility and crop productivity.

Source: OECD/Eurostat (2012), *Nitrogen and Phosphorus Balance Handbook*, *www.oecd.org/tad/sustainable-agriculture/agri-environmentalindicators.htm*.

pressures, such as declining soil fertility in the case of a nutrient deficit, or for a nutrient surplus the risk of polluting soil, water and air.

The nutrient balances are expressed in terms of changes in the physical quantities of nutrient surpluses (deficits) to indicate the trend and level of potential physical pressure of nutrient surpluses into the environment. The nutrient balance indicator is also expressed in terms of kilogrammes of nutrient surplus (deficit) per hectare of agricultural land per annum to facilitate the comparison of the relative intensity of nutrients in agricultural systems between countries.

When interpreting these indicators it should be noted that they describe potential environmental pressures, and may hide important sub-national variations. More information is needed to describe the actual pressure. They should be read together with information on agricultural land use and farm management approaches. Cross-country comparisons of change in nutrient surplus intensities over time should take into account the absolute intensity levels during the reference period. It should also be noted that these indicators reflect nutrient balances from primary agriculture, and do not consider nutrient flows from other food production systems, such as fisheries or total nitrogen cycles in the economy.

Limitations of nutrient balances include the accuracy of the underlying nutrient conversion coefficients and also the uncertainties involved in estimating nutrient uptake by areas of pasture and some fodder crops. In addition, environmental events like droughts and floods will affect the efficiency of plants to fix nutrients. The soil science of nutrients is also not well understood (e.g. soils vary in their capacity to store nutrients) and there is limited information on the varietal mix of legumes in pastures to accurately estimate pasture uptake of nitrogen (OECD, 2008).

As an environmental *driving force*, nutrient balance indicators link to the *state* of nutrients in water (Chapter 9), emissions of ammonia (Chapter 10) and greenhouse gas emissions (Chapter 11).

Measurability and data quality

OECD and Eurostat data on nitrogen and phosphorus balances are available for all OECD countries, except Chile, until 2009 (Annex 1.A2). Improvements to the underlying methodology, nutrient conversion coefficients and primary data are currently being undertaken by OECD countries in cooperation with Eurostat, as the nutrient balances are revised and updated. For example, Eurostat is examining how to account for biological nitrogen fixation by clover in pasture.

4.3. Main trends

Overall OECD agricultural nutrient surpluses (N and P) have been on a continuous downward trend from 1990 to 2009, both in absolute tonnes of nutrients (Figures 4.2 and 4.3) and in terms of nutrient surpluses per hectare of agricultural land (Figures 4.4 and 4.5). The rate of reduction in OECD nutrient surpluses was more rapid over the 2000s compared to the 1990s.

The lowering of nutrient surpluses has reduced the risk of environmental pressure on soil, water and air. This reflects both overall improvements in nutrient use efficiency by farmers, and the slower growth in agricultural production for many countries over the 2000s (Figure 3.1). There are, however, sizeable variations within and across countries in terms of the intensity and trends of nutrient surpluses.

Despite the overall improvement in lowering nutrient surpluses, nitrogen (N) and phosphorus (P) intensity levels per hectare of agricultural land remain at very high levels in terms of their potential to cause environmental damage. ***Background (or natural) loss*** of N is typically estimated at around 1-2 kg/ha from electrical storms and other sources, while for phosphorus this figure is about 0.1 kg/ha depending on underlying conditions in sediment and rocks (OECD, 2012).

By 2007-09, around two thirds of OECD countries had an annual national nitrogen surplus in excess of 40 kgN/ha nitrogen, with **Belgium**, **Israel**, **Japan**, **Korea** and the **Netherlands** with a surplus in excess of 100 kgN/ha (Figure 4.3). Similarly for phosphorus about a third of OECD countries have a surplus in excess of 5 kgP/ha, with **Israel**, **Japan**, **Korea**, the **Netherlands**, and **Norway**, with a surplus in excess of 10 kgP/ha (Figure 4.5).

Some countries (**Estonia**, **Greece**, **Hungary**, **Italy** and the **Slovak Republic**) experienced an absolute phosphorus deficit in 2007-09 (Figures 4.3 and 4.5). While if prolonged this phosphorus deficit could undermine soil fertility (a possibility in the case of **Hungary** which has experienced a 20 year P deficit), it is likely that in most of these cases crops and pasture can draw on P soil stores accumulated over many years of previous over application of P to soils.

For the few OECD countries where nutrient surpluses have been growing over the past decade, these countries frequently have levels of surpluses expressed in terms of N or P per hectare of agricultural land below the OECD average, such as for **Canada**, **New Zealand** (not for phosphorus) and **Poland**. For **Israel**, however, nutrient surplus in terms of kg of N/P per hectare are appreciably higher than the OECD average and grew over the past 10 years.

The OECD average reduction in P surpluses, both in absolute terms and expressed in kg of P per hectare, has been more than double the rate of reduction per annum compared to N surpluses over the past 20 years (Figures 4.2 to 4.5). To a large extent, especially over the past decade, this reflects the realisation by farmers that their soils had high levels of accumulated phosphorus from which crops and pasture can draw without further

Figure 4.2. **Nitrogen balance volume, OECD countries, 1990-2009**

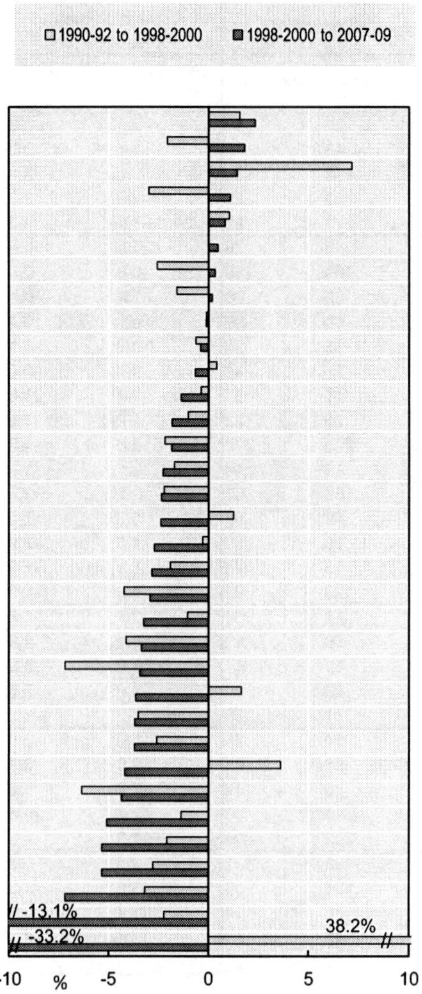

	Average (Thousand tonnes of nitrogen)			Average annual % change	
	1990-92[1]	1998-2000[2]	2007-09[3]	1990-92 to 1998-2000	1998-2000 to 2007-09
New Zealand	399	451	554	1.5	2.3
Poland	896	757	889	-2.1	1.8
Canada	753	1 311	1 489	7.2	1.4
Czech Republic	327	256	282	-3.0	1.1
Iceland	17	19	20	1.0	0.8
Israel	..	53	55	..	0.5
Switzerland	122	99	102	-2.6	0.3
Japan	935	821	834	-1.6	0.2
Norway	103	102	101	-0.1	-0.1
Turkey	1 493	1 418	1 367	-0.6	-0.4
United States	13 930	14 378	13 534	0.4	-0.7
OECD[4]	**40 052**	**38 869**	**34 270**	**-0.4**	**-1.4**
Mexico	2 885	2 655	2 286	-1.0	-1.9
France	2 100	1 757	1 510	-2.2	-1.9
United Kingdom	1 696	1 424	1 182	-1.7	-2.3
Germany	2 131	1 778	1 432	-2.2	-2.4
Korea	465	513	412	1.2	-2.4
Australia	7 527	7 353	5 730	-0.3	-2.7
EU15	**9 966**	**8 529**	**6 567**	**-1.9**	**-2.9**
Slovenia	50	38	30	-4.2	-3.0
Sweden	197	180	134	-1.1	-3.3
Denmark	460	328	241	-4.1	-3.4
Slovak Republic	177	97	71	-7.2	-3.5
Spain	627	713	510	1.6	-3.7
Finland	204	153	109	-3.5	-3.7
Italy	667	540	398	-2.6	-3.7
Ireland	236	313	213	3.6	-4.2
Greece	309	183	128	-6.4	-4.4
Netherlands	655	586	386	-1.4	-5.1
Belgium	306	258	167	-2.1	-5.3
Austria	195	156	95	-2.8	-5.4
Luxembourg	23	18	10	-3.2	-7.2
Portugal	160	143	54	-2.2	-13.1
Hungary	5	71	3	38.2	-33.2
Estonia	..	..	17	..	..

..: not available.

Notes: The gross nitrogen balance (surplus or deficit) calculates the difference between the nitrogen inputs entering a farming system (i.e. mainly livestock manure and fertilisers) and the nitrogen outputs leaving the system (i.e. the uptake of nitrogen for crop and pasture production).

Countries are ranked in descending order according to average annual percentage change 1998-2000 to 2007-09.

The statistical data for Israel are supplied by and under the responsibility of the relevant Israeli authorities. The use of such data by the OECD is without prejudice to the status of the Golan Heights, East Jerusalem and Israeli settlements in the West Bank under the terms of international law.

1. Data for 1990-92 average refer to the year 1990 for the United Kingdom; the 1992-94 average for Slovenia; and the 1995-97 average for Portugal.
2. Data for 1998-2000 average refer to the year 2000 for the United Kingdom; and the 2000-02 average for Israel and Portugal.
3. Data for 2007-09 average refer to the 2006-08 average for Belgium, France, Greece, Hungary, Italy, Luxembourg, Mexico, Netherlands, Slovenia and Switzerland.
4. The OECD total excludes Chile, Estonia and Israel.

Source: OECD/Eurostat Agri-Environmental Indicator Database, http://epp.eurostat.ec.europa.eu; and national data for Spain.

StatLink *http://dx.doi.org/10.1787/888932792635*

applications of P, at least for a number of years. The understanding that agricultural soils have high stocks of P, has come from, in particular, improved methods and frequency of soil nutrient testing by farmers in many OECD countries (discussed further below).

An encouraging development in a growing number of countries has been the ***decoupling of the growth in agricultural production from changes in nutrient surpluses***. Environmental decoupling occurs when the relative growth rate of the environmentally relevant variable (i.e. here nutrient surpluses) is less than the growth rate of the variable reflecting the economic driving force (i.e. here agricultural production). Between 1998-2000

Figure 4.3. **Nitrogen balance per hectare of agricultural land, OECD countries, 1990-2009**

□ 1990-92 to 1998-2000 ■ 1998-2000 to 2007-09

-12.9% -30.3% 36.4%

-10 % -5 0 5 10

	Average (kg nitrogen/ha)			Average annual % change	
	1990-92[1]	1998-2000[2]	2007-09[3]	1990-92 to 1998-2000	1998-2000 to 2007-09
New Zealand	30	35	49	2.0	3.6
Poland	49	43	57	-1.7	3.3
Israel	..	103	130	..	3.3
Czech Republic	79	63	79	-2.9	2.6
Canada	12	21	23	7.1	1.3
Iceland	7	8	8	1.1	0.9
Japan	180	169	180	-0.8	0.7
Switzerland[4]	80	65	68	-2.4	0.5
Norway	103	98	99	-0.6	0.1
Turkey	37	36	35	-0.2	-0.4
United States	33	35	33	0.7	-0.6
Slovak Republic	76	41	37	-7.4	-1.0
Mexico	28	25	22	-1.2	-1.4
Korea	213	263	228	2.6	-1.6
Australia	16	16	14	-0.1	-1.7
Germany	123	103	85	-2.2	-2.2
France	69	59	50	-1.9	-2.2
OECD[5]	**86**	**78**	**63**	**-1.3**	**-2.2**
Italy	38	35	28	-1.1	-2.5
United Kingdom	140	120	97	-1.5	-2.6
Slovenia	90	77	61	-2.7	-2.8
Sweden	58	56	43	-0.4	-2.9
Spain	16	19	14	1.9	-3.5
Denmark	166	123	90	-3.6	-3.5
Ireland	53	71	50	3.7	-3.7
Greece	60	34	25	-6.9	-3.8
EU15	**109**	**93**	**65**	**-2.0**	**-3.9**
Finland	80	70	47	-1.8	-4.2
Austria	56	46	30	-2.6	-4.6
Netherlands	331	302	204	-1.2	-4.8
Belgium	227	186	121	-2.5	-5.2
Luxembourg	183	137	75	-3.6	-7.3
Portugal	41	38	14	-1.7	-12.9
Hungary	1	12	1	36.4	-30.3
Estonia	..	..	21	..	..

..: not available.

Notes: Balance (surplus or deficit) expressed as kg nitrogen per hectare of total agricultural land.

Countries are ranked in descending order according to average annual percentage change 1998-2000 to 2007-09.

The statistical data for Israel are supplied by and under the responsibility of the relevant Israeli authorities. The use of such data by the OECD is without prejudice to the status of the Golan Heights, East Jerusalem and Israeli settlements in the West Bank under the terms of international law.

1. Data for 1990-92 average refer to the year 1990 for the United Kingdom; the 1992-94 average for Slovenia; and the 1995-97 average for Portugal.
2. Data for 1998-2000 average refer to the year 2000 for the United Kingdom; and the 2000-02 average for Israel and Portugal.
3. Data for 2007-09 average refer to the 2006-08 average for Belgium, France, Greece, Hungary, Italy, Luxembourg, Mexico, Netherlands, Slovenia and Switzerland.
4. In the case of Switzerland, total agricultural area includes summer grazing.
5. The OECD total excludes Chile, Estonia and Israel.

Source: OECD/Eurostat Agri-Environmental Indicator Database, http://epp.eurostat.ec.europa.eu; and national data for Spain.

StatLink http://dx.doi.org/10.1787/888932792654

and 2007-09, while overall OECD volume of agricultural production increased by more than 1% per annum, the nitrogen balance (tonnes) declined by over 1% per annum, while the phosphorus balance (tonnes) decreased by over 5% per annum (Figures 4.6 and 4.7).

Much of this improvement has resulted from the increasing adoption of nutrient management practices encouraged by extensive agri-environmental measures across many countries. The much greater decoupling for phosphorus compared to nitrogen is largely explained by the growing realisation by farmers in many OECD countries of the accumulation of P in soils, especially with the more widespread use of soil nutrient testing.

Figure 4.4. **Phosphorus balance volume, OECD countries, 1990-2009**

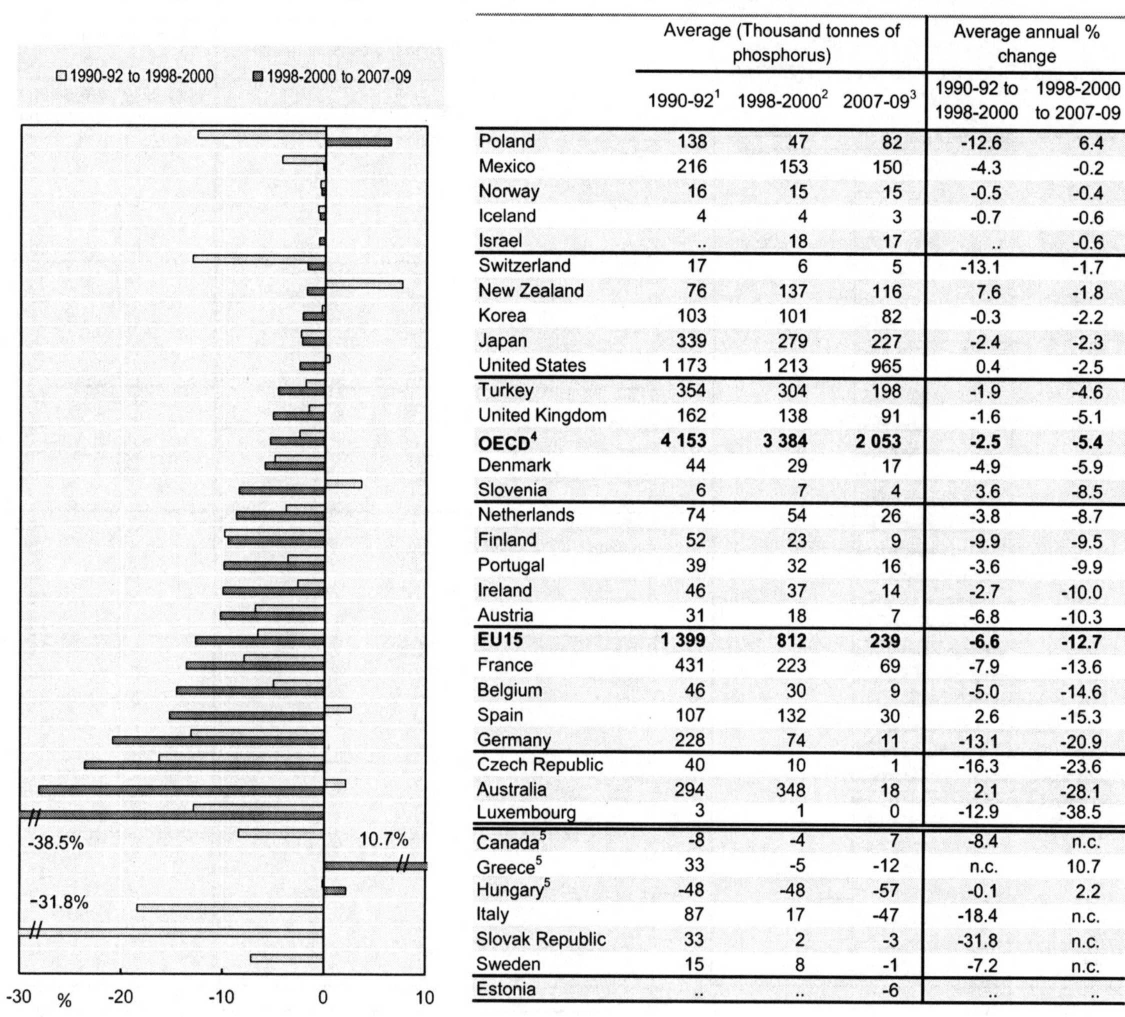

	Average (Thousand tonnes of phosphorus)			Average annual % change	
	1990-92[1]	1998-2000[2]	2007-09[3]	1990-92 to 1998-2000	1998-2000 to 2007-09
Poland	138	47	82	-12.6	6.4
Mexico	216	153	150	-4.3	-0.2
Norway	16	15	15	-0.5	-0.4
Iceland	4	4	3	-0.7	-0.6
Israel	..	18	17	..	-0.6
Switzerland	17	6	5	-13.1	-1.7
New Zealand	76	137	116	7.6	-1.8
Korea	103	101	82	-0.3	-2.2
Japan	339	279	227	-2.4	-2.3
United States	1 173	1 213	965	0.4	-2.5
Turkey	354	304	198	-1.9	-4.6
United Kingdom	162	138	91	-1.6	-5.1
OECD[4]	**4 153**	**3 384**	**2 053**	**-2.5**	**-5.4**
Denmark	44	29	17	-4.9	-5.9
Slovenia	6	7	4	3.6	-8.5
Netherlands	74	54	26	-3.8	-8.7
Finland	52	23	9	-9.9	-9.5
Portugal	39	32	16	-3.6	-9.9
Ireland	46	37	14	-2.7	-10.0
Austria	31	18	7	-6.8	-10.3
EU15	**1 399**	**812**	**239**	**-6.6**	**-12.7**
France	431	223	69	-7.9	-13.6
Belgium	46	30	9	-5.0	-14.6
Spain	107	132	30	2.6	-15.3
Germany	228	74	11	-13.1	-20.9
Czech Republic	40	10	1	-16.3	-23.6
Australia	294	348	18	2.1	-28.1
Luxembourg	3	1	0	-12.9	-38.5
Canada[5]	-8	-4	7	-8.4	n.c.
Greece[5]	33	-5	-12	n.c.	10.7
Hungary[5]	-48	-48	-57	-0.1	2.2
Italy	87	17	-47	-18.4	n.c.
Slovak Republic	33	2	-3	-31.8	n.c.
Sweden	15	8	-1	-7.2	n.c.
Estonia	..	..	-6	..	..

..: not available.

n.c.: not calculated.

Notes: The gross phosphorus balance (surplus or deficit) calculates the difference between the phosphorus inputs entering a farming system (i.e. mainly livestock manure and fertilisers) and the phosphorus outputs leaving the system (i.e. the uptake of phosphorous for crop and pasture production). Countries are ranked in descending order according to average annual percentage change 1998-2000 to 2007-09.

The statistical data for Israel are supplied by and under the responsibility of the relevant Israeli authorities. The use of such data by the OECD is without prejudice to the status of the Golan Heights, East Jerusalem and Israeli settlements in the West Bank under the terms of international law.

1. Data for 1990-92 average refer to the year 1990 for the United Kingdom; and the 1995-97 average for Portugal and Slovenia.
2. Data for 1998-2000 average refer to the year 2000 for the United Kingdom; and the 2000-02 average for Israel, Portugal and Slovenia.
3. Data for 2007-09 average refer to the 2006-08 average for Belgium, Estonia, France, Germany, Greece, Hungary, Italy, Luxembourg, Mexico, Netherlands, Slovenia and Switzerland.
4. The OECD total excludes Chile, Estonia and Israel.
5. For Canada, Greece and Hungary, the average annual percentage change refers to change in phosphorus deficit.

Source: OECD/Eurostat Agri-Environmental Indicator Database, http://epp.eurostat.ec.europa.eu; and national data for Spain.

StatLink http://dx.doi.org/10.1787/888932792673

Moreover, gains in P use efficiency have also been achieved through changing livestock husbandry practices, especially by altering animal feed dietary composition (OECD, 2008).

The physical properties of P in the environment are different compared to N, but ***the accumulation of P in farm soils beyond crop needs in many OECD countries is a growing environmental concern*** (OECD, 2008). The retention of particulate P in soils is generally high compared to N, hence, it is usually transported with long time lags into surface water

Figure 4.5. **Phosphorus balance per hectare of agricultural land, OECD countries, 1990-2009**

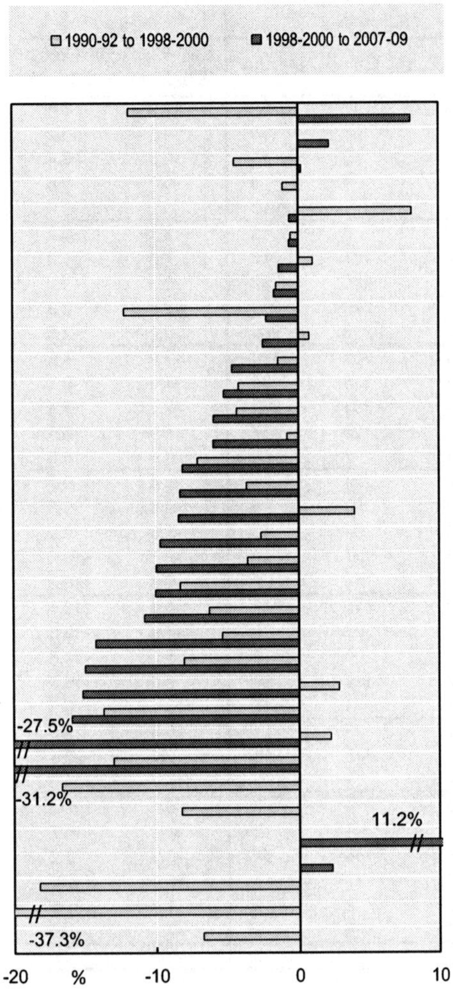

□ 1990-92 to 1998-2000 ■ 1998-2000 to 2007-09

-27.5% -31.2% 11.2% -37.3%

-20 % -10 0 10

	Average (kg phosphorus/ha)			Average annual % change	
	1990-92[1]	1998-2000[2]	2007-09[3]	1990-92 to 1998-2000	1998-2000 to 2007-09
Poland	7	3	5	-11.9	8.0
Israel	..	34	40	..	2.2
Mexico	2	1	1	-4.5	0.2
Norway	16	15	15	-1.1	0.0
New Zealand	6	11	10	8.1	-0.6
Iceland	2	2	1	-0.5	-0.7
Korea	47	52	46	1.0	-1.4
Japan	65	57	49	-1.6	-1.7
Switzerland[4]	11	4	3	-12.2	-2.3
United States	3	3	2	0.8	-2.5
Turkey	9	8	5	-1.4	-4.6
OECD[5]	**13**	**9**	**6**	**-4.2**	**-5.2**
Denmark	16	11	6	-4.3	-5.9
United Kingdom	13	12	7	-0.8	-6.0
Austria	9	5	2	-7.1	-8.1
Netherlands	38	28	14	-3.6	-8.3
Slovenia	12	15	9	4.1	-8.4
Ireland	10	8	3	-2.7	-9.7
Portugal	10	8	4	-3.6	-10.0
Finland	21	10	4	-8.3	-10.0
EU15	**14**	**9**	**3**	**-6.3**	**-10.8**
Belgium	34	22	6	-5.4	-14.3
France	14	7	2	-8.0	-15.0
Spain	3	3	1	2.9	-15.1
Germany	13	4	1	-13.7	-15.9
Australia	1	1	0	2.3	-27.5
Luxembourg	20	7	0	-13.0	-31.2
Czech Republic	10	2	0	-16.6	n.c.
Canada[6]	0	0	0	-8.2	n.c.
Greece[6]	6	-1	-2	n.c.	11.2
Hungary[6]	-8	-8	-10	0.0	2.4
Italy	5	1	-3	-18.2	n.c.
Slovak Republic	14	0	-2	-37.3	n.c.
Sweden	5	3	0	-6.8	n.c.
Estonia	..	..	-7	..	..

..: not available.

n.c.: not calculated.

Notes: Countries are ranked in descending order according to average annual percentage change 1998-2000 to 2007-09.

Balance (surplus or deficit) expressed as kg phosphorus per hectare of total agricultural land.

The statistical data for Israel are supplied by and under the responsibility of the relevant Israeli authorities. The use of such data by the OECD is without prejudice to the status of the Golan Heights, East Jerusalem and Israeli settlements in the West Bank under the terms of international law.

1. Data for 1990-92 average refer to the year 1990 for the United Kingdom; and the 1995-97 average for Portugal and Slovenia.
2. Data for 1998-2000 average refer to the year 2000 for the United Kingdom; and the 2000-02 average for Israel, Portugal and Slovenia.
3. Data for 2007-09 average refer to the 2006-08 average for Belgium, Estonia, France, Germany, Greece, Hungary, Italy, Luxembourg, Mexico, Netherlands, Slovenia and Switzerland.
4. In the case of Switzerland, total agricultural area includes summer grazing.
5. The OECD total excludes Chile, Estonia and Israel.
6. For Canada, Greece and Hungary, the average annual percentage change refers to change in phosphorus deficit.

Source: OECD/Eurostat Agri-Environmental Indicator Database, http://epp.eurostat.ec.europa.eu; and national data for Spain.

StatLink *http://dx.doi.org/10.1787/888932792692*

through soil erosion rather than leaching into groundwater, unlike the more rapid transport of N from soils into water bodies. Hence, it is likely that there will be a considerable time lag for many countries between reductions in P surpluses leading to lower P concentrations in water supplies. Indeed, P concentrations in rivers and lakes could continue to rise over decades into the future, while the implications for groundwater are unclear (OECD, 2012).

Figure 4.6. Nitrogen balance and agricultural production volume, OECD countries, 1990-2009

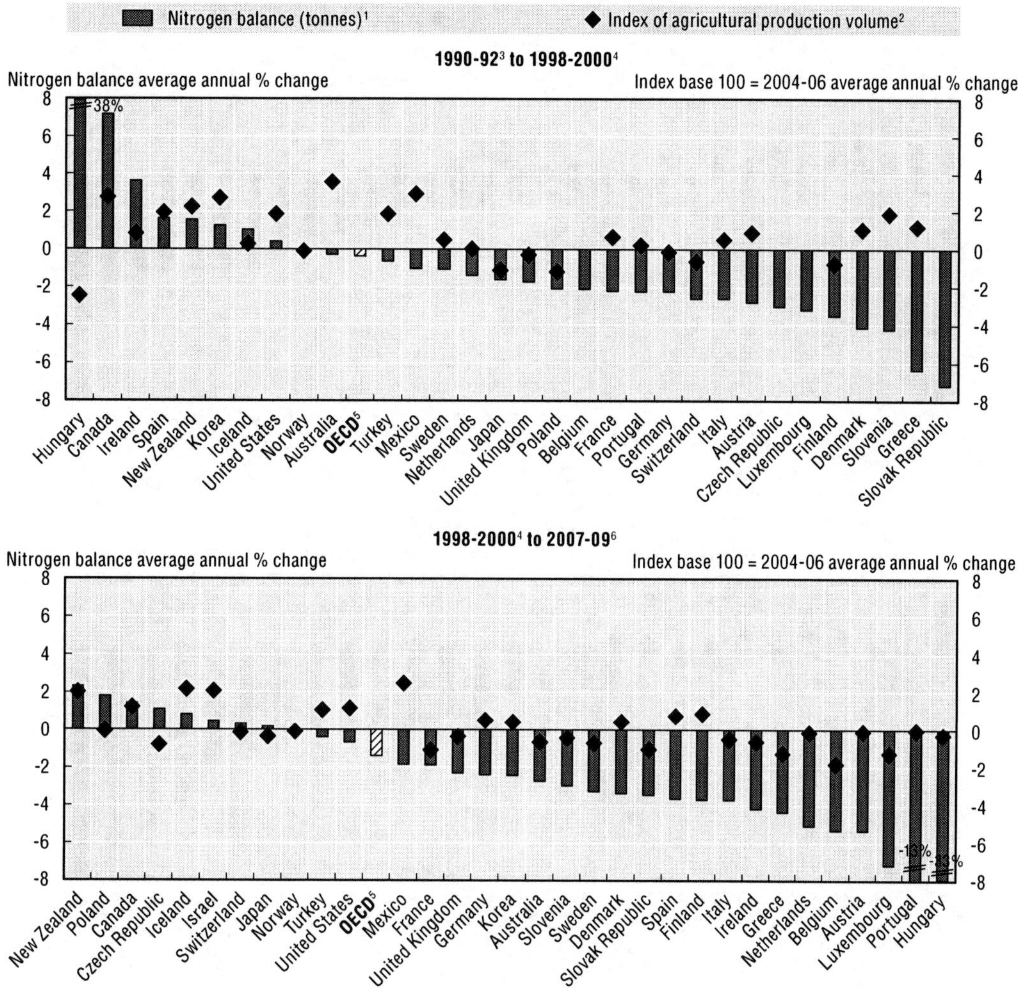

Nitrogen balance (tonnes)[1] ◆ Index of agricultural production volume[2]

1990-92[3] to 1998-2000[4]

Nitrogen balance average annual % change — Index base 100 = 2004-06 average annual % change

38%

Hungary, Canada, Ireland, Spain, New Zealand, Korea, Iceland, United States, Norway, Australia, OECD[5], Turkey, Mexico, Sweden, Netherlands, Japan, United Kingdom, Poland, Belgium, France, Portugal, Germany, Switzerland, Italy, Austria, Czech Republic, Luxembourg, Finland, Denmark, Slovenia, Greece, Slovak Republic

1998-2000[4] to 2007-09[6]

Nitrogen balance average annual % change — Index base 100 = 2004-06 average annual % change

-13% -13%

New Zealand, Poland, Canada, Czech Republic, Iceland, Israel, Switzerland, Japan, Norway, Turkey, United States, OECD[5], Mexico, France, United Kingdom, Germany, Korea, Australia, Slovenia, Sweden, Denmark, Slovak Republic, Spain, Finland, Italy, Ireland, Greece, Netherlands, Belgium, Austria, Luxembourg, Portugal, Hungary

Notes: Countries are ranked in descending order according to nitrogen balance average annual percentage change.

The statistical data for Israel are supplied by and under the responsibility of the relevant Israeli authorities. The use of such data by the OECD is without prejudice to the status of the Golan Heights, East Jerusalem and Israeli settlements in the West Bank under the terms of international law.

1. The gross nitrogen balance (surplus or deficit) calculates the difference between the nitrogen inputs entering a farming system (i.e. mainly livestock manure and fertilisers) and the nitrogen outputs leaving the system (i.e. the uptake of nitrogen for crop and pasture production).
2. The FAO indices of agricultural production show the relative level of the aggregate volume of agricultural production for each year in comparison with the base period 2004-06. They are based on the sum of price weighted quantities of different agricultural commodities produced after deductions of quantities used as seed and feed weighted in a similar manner. The resulting aggregate represents, therefore, disposable production for any use except as seed and feed. All the indices at the country, regional and world levels are calculated by the Laspeyres formula. Production quantities of each commodity are weighted by 2004-06 average international commodity prices and summed for each year. To obtain the index, the aggregate for a given year is divided by the average aggregate for the base period 2004-06.
3. Data for 1990-92 average refer to the year 1990 for the United Kingdom; the 1992-94 average for Slovenia; and the 1995-97 average for Portugal.
4. Data for 1998-2000 average refer to the year 2000 for the United Kingdom; and the 2000-02 average for Israel and Portugal.
5. The OECD total excludes Chile, Estonia and Israel.
6. Data for 2007-09 average refer to the 2006-08 average for Belgium, France, Greece, Hungary, Italy, Luxembourg, Mexico, Netherlands, Slovenia and Switzerland.

Source: FAOSTAT (2012), "Food and Agriculture Organisation of the United Nations", *http://faostat.fao.org; OECD/Eurostat Agri-Environmental Indicator Database, http://epp.eurostat.ec.europa.eu;* and national data for Spain.

StatLink http://dx.doi.org/10.1787/888932792711

Figure 4.7. **Phosphorus balance and agricultural production volume, OECD countries, 1990-2009**

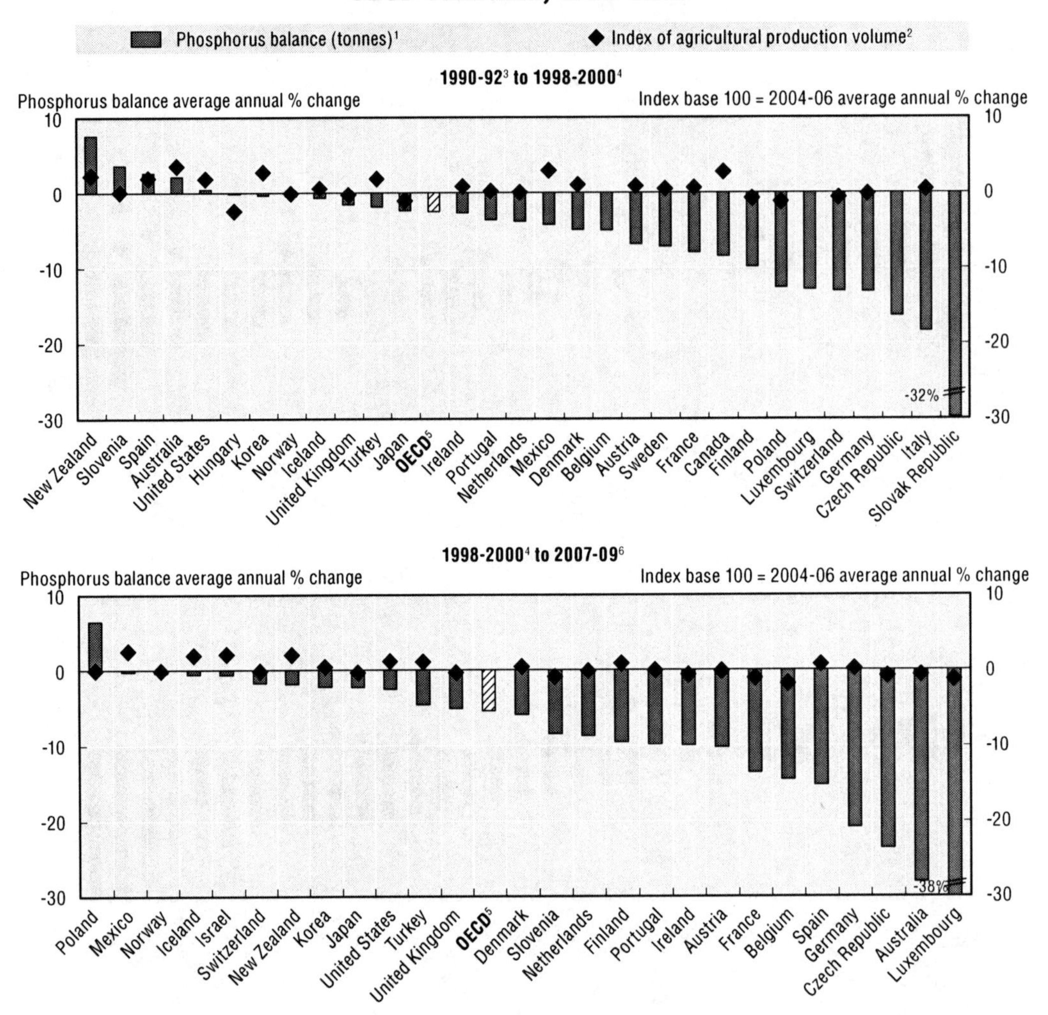

Notes: Countries are ranked in descending order according to phosphorus balance average annual percentage change.

The statistical data for Israel are supplied by and under the responsibility of the relevant Israeli authorities. The use of such data by the OECD is without prejudice to the status of the Golan Heights, East Jerusalem and Israeli settlements in the West Bank under the terms of international law.

1. The gross phosphorus balance (surplus or deficit) calculates the difference between the phosphorus inputs entering a farming system (i.e. mainly livestock manure and fertilisers) and the phosphorus outputs leaving the system (i.e. the uptake of phosphorus for crop and pasture production). Countries with a phosphorus deficit are not included in the figure, as follows: Canada, Estonia, Greece, Hungary, Italy (1998-2000 to 2007-09), Slovak Republic (1998-2000 to 2007-09) and Sweden (1998-2000 to 2007-09).
2. The FAO indices of agricultural production show the relative level of the aggregate volume of agricultural production for each year in comparison with the base period 2004-06. They are based on the sum of price weighted quantities of different agricultural commodities produced after deductions of quantities used as seed and feed weighted in a similar manner. The resulting aggregate represents, therefore, disposable production for any use except as seed and feed. All the indices at the country, regional and world levels are calculated by the Laspeyres formula. Production quantities of each commodity are weighted by 2004-06 average international commodity prices and summed for each year. To obtain the index, the aggregate for a given year is divided by the average aggregate for the base period 2004-06.
3. Data for 1990-92 average refer to the year 1990 for the United Kingdom; and the 1995-97 average for Portugal and Slovenia.
4. Data for 1998-2000 average refer to the year 2000 for the United Kingdom; and the 2000-02 average for Israel, Portugal and Slovenia.
5. The OECD total excludes Chile, Estonia and Israel.
6. Data for 2007-09 average refer to the 2006-08 average for Belgium, France, Germany, Greece, Hungary, Italy, Luxembourg, Mexico, Netherlands, Slovenia and Switzerland.

Source: FAOSTAT (2012), *Food and Agriculture Organisation of the United Nations, http://faostat.fao.org;* OECD/Eurostat *Agri-Environmental Indicator Database, http://epp.eurostat.ec.europa.eu;* and national data for Spain.

StatLink http://dx.doi.org/10.1787/888932792730

Figure 4.8. **Spatial distribution of nitrogen balances, Canada and Poland, 1991-2009**

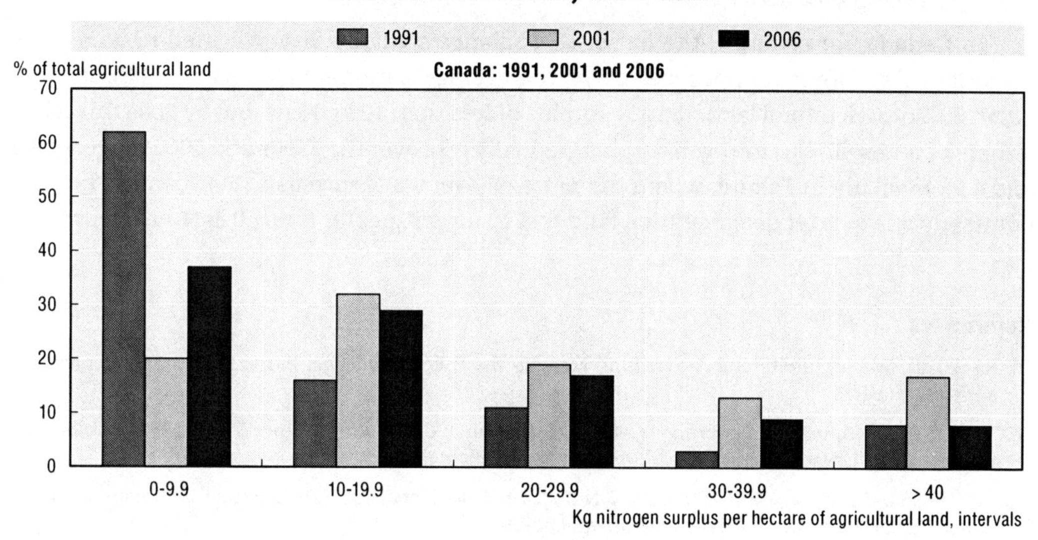

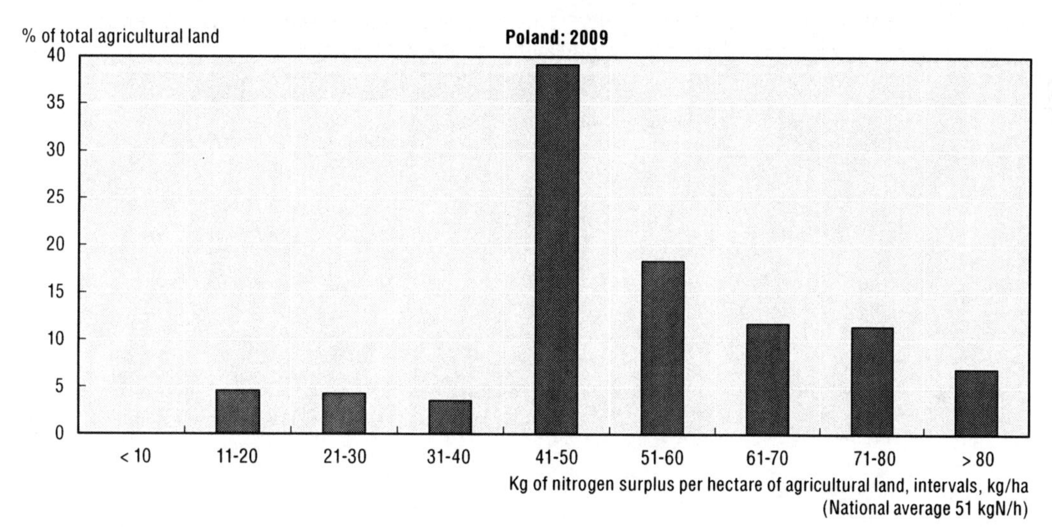

Source: Panel A: Eilers, W., R. MacKay, L. Graham and A. Lefebvre (eds.) (2010), "Environmental sustainability of Canadian agriculture", *Agri-environmental indicator report series*, Report #3, Agriculture and Agri-Food Canada, Ottawa, Canada, *http://publications.gc.ca/collections/collection_2011/agr/A22-201-2010-eng.pdf*. Panel B: Polish Ministry of Agriculture and Rural Development.

StatLink http://dx.doi.org/10.1787/888932792749

In most countries there is considerable variation in the level and trends of ***regional nutrient surpluses*** around national average values. Regional variations are explained by the spatial distribution of intensive livestock farming and cropping systems that require high nutrient inputs, such as maize and rice relative to wheat and oilseeds, as well as differing climates and types of soil, and also varying topography across the agricultural regions.

National nutrient balance indicators can mask important regional (sub-national) variations across a country, especially where more intensive agricultural production systems are spatially concentrated in a small part of the overall agricultural land area. While **Australia**, **Canada**, **Mexico**, and the **United States,** for example, have nutrient surplus intensities below the OECD average (expressed as kgN/P/ha of agricultural land, Figures 4.3 and 4.5) there are regions within these countries where excess nutrients place a considerable burden on the environment or where nutrient deficits have potential to undermine crop productivity.

In **Canada**, for example, the national N balance spatially disaggregated reveals some important developments not revealed by the average national value (Figure 4.8). In 1991, about 60% of agricultural land had a N surplus of less than 10 kg/N/ha, but by 2006 this fell to a share of under 40%, as nitrogen surplus gradually rose over the 1990s and 2000s (Figures 4.2 and 4.3). Similarly in **Poland**, where the national average N surplus in 2009 was 51 kgN/ha, but nearly one quarter of agricultural land had a surplus greater than 60 kgN/ha (Figure 4.8).

References

OECD (2012), *Water Quality and Agriculture: Meeting the Policy Challenge*, Paris, France, *www.oecd.org/agriculture/water*.

OECD (2008), *Environmental Performance of Agriculture in OECD Countries Since 1990*, OECD Publishing, *www.oecd.org/tad/sustainable-agriculture/agri-environmentalindicators.htm*.

OECD/EUROSTAT (2012a), *OECD/EUROSTAT Nitrogen Balance Handbook*, OECD Publishing, *www.oecd.org/tad/sustainable-agriculture/agri-environmentalindicators.htm*.

OECD/EUROSTAT (2012b), *OECD/EUROSTAT Phosphorus Balance Handbook*, OECD Publishing, *www.oecd.org/tad/sustainable-agriculture/agri-environmentalindicators.htm*.

Chapter 5

Pesticide sales

This chapter reviews the environmental performances of agriculture in OECD countries related to pesticide sales. It provides a description of the policy context (issues and main challenges), definitions for the agri-environmental indicators presented, and elements related to concepts, interpretations, links to other indicators, as well as measurability and data quality. The chapter then describes the main trends of the agri-environmental indicators, using available data covering the period 1990-2010 and based on a set of tables and figures.

5.1. Policy context

The issue

Pesticides are major inputs for agriculture that facilitate lowering the risks of yield losses. As agriculture is the major user of pesticides it is also a significant source of risk of pollution into water systems and of concern for human and wildlife health and the functioning of ecosystems. This concerns all OECD countries, and as a result there is an extensive range of policy instruments used by countries to address human and ecosystem health concerns and pesticide pollution of water (Chapter 9; and OECD, 2012) and air in terms of the sales of methyl bromide (Chapter 12), notably: regulatory instruments (e.g. human health and environmental risk assessment prior to marketing and the sale of pesticides); payments to encourage adoption of practices that lower use and lead to more accurate application; pesticide taxes to encourage greater use efficiency by farmers; and farm advice and information.

Main challenges

The main challenge is to reduce the risks to human health, ecosystems and water systems from excessive exposure to pesticides, while maintaining and increasing the level of crop productivity. This requires taking into account the different factors affecting pesticide risks in the environment, for example, the handling and storage of pesticides on-farms, the toxicity and persistence of pesticides in the environment, and weather conditions during the field application of pesticides.

5.2. Indicators

Definitions

The indicator related to agricultural pesticide sales includes the change in:

- Pesticide sales, in tonnes of active ingredients.

Concepts, interpretation, limitations and links to other indicators

There exist different types of pesticides according to their chemical composition and their targets: biocides, insecticides, fungicides, etc. Pesticide sales data are a proxy measure of potential environmental pressure, since it does not convey information on the real levels of risk exposures for ecosystems and human health, which depend on other factors including toxicity, mobility and persistence. The indicator of pesticide sales tracks trends over time in the overall quantity purchased by agriculture (data refer to sales of active ingredients of insecticides, fungicides, herbicides and other pesticides including plant growth regulators and rodenticides).

Care is required when comparing absolute levels of pesticide sales across countries, because of differences in climatic conditions and farming systems, which affect the composition and level of usage (OECD, 2008). Variability of climatic conditions (especially temperature and precipitation), may markedly alter annual pesticide use, while changes in the mix of pesticides can reduce active ingredients applied but increase adverse impacts. The indicator does not recognise the differences among pesticides in their levels of toxicity, persistence and mobility. In addition, the greater use by farmers of pesticides with lower potential risk to humans and the environment because they are more narrowly targeted or degrade more rapidly, might not be revealed by any change in overall pesticide sale trends, and possibly even show an increase.

This indicator is not expressed in terms of the quantity of pesticide sales per hectare of agricultural land (or crop land), unlike that which commonly appears in many studies. This is because the application of pesticides varies widely for different crops, both within and across countries, and is sometimes used in the cultivation of forage crops, but limited cross country time series data exist in this regard (OECD, 2008). A limitation in the use of the indicator as a comparative index across countries is that the definition and coverage of pesticide sales data vary across OECD countries, as discussed in the following section.

At present, despite the limitations of assessing pesticide impacts on human health and the environment with the pesticide sales indicator, there are no alternative indicators currently available. A few OECD countries have developed risks indicators, but despite many years of international effort by OECD and other organisations to develop a harmonised and comparable set of pesticide risks the goal remains elusive (OECD, 2008). Even so, the pesticide sales indicator has been used in a policy context by some OECD countries, for example, where pesticide taxes have been introduced or in pesticide reduction plans, such as the French *Ecophyto* plan which aims to reduce pesticide sales by 50% in France by 2018 (OECD, 2012). In addition, this indicator is now widely used by a diverse group of national governments, international governmental organisations and non-governmental organisations, in their regular reporting and monitoring of environmental trends.

As an environmental driving force, the pesticide sales indicator links to the *state* (or concentration) of pesticides in water bodies (Chapter 9), and emissions of methyl bromide which has the potential to deplete the ozone layer (Chapter 12).

Measurability and data quality

In most OECD countries, the available data refers to pesticide sales, which provides an imperfect proxy of agricultural pesticide use or consumption on farms. This is because, first, pesticide sales may be different from pesticide use because of farmers' storage of pesticides. Second, pesticide sales not only cover the agricultural sector but also other activities such as sales for urban use (e.g. road and rail verges), private gardens, golf courses and forestry. For example, in the **United States**, the total amount of pesticide active ingredient sales have been estimated to be shared between 80% purchased by agriculture, 12% by industry, commercial and government purposes, and 8% by home and garden owners. In **Belgium**, agriculture accounts for about 65-70% of sales (EPA, 2011; OECD, 2008).

For all countries, the data represent pesticide sales, except for **Korea** and **Mexico** which are national production data and the **United Kingdom** where data concern the amount of active substance applied on-farm, i.e. usage. Pesticide sales covers agriculture and non-agricultural sales (e.g. forestry, gardens), except for the following countries which only include agriculture: **Belgium**, **Denmark** and **Sweden**. Data are not available for **Israel**, while **Luxembourg** is included in Belgium.

5.3. Main trends

Overall OECD pesticide sales diminished by -1.1% per annum over the period 2000-10, which contrasts to a small per annum increase over the 1990s of +0.2% per annum (Figure 5.1). Much of the declining sales of pesticides over the last decade was accounted for by the **EU15** and the **United States**, which together accounted for 70% of total pesticide sales by 2008-10 (Figure 5.1). A number of other major users of pesticides across OECD also experienced a reduction or no change in pesticide sales over the most recent period, including **France**, **Italy**, and **Japan** (Figures 5.1 and 5.2).

Nearly all EU transition economies (notably **Estonia**, the **Czech and Slovak Republics** and **Hungary**, revealed a strong growth in pesticide purchases over the 2000s, compared to the 1990s when in many cases sales declined (Figures 5.1 and 5.2). Following the reductions in pesticide sales for these countries in their move toward a market economy over the 1990s, the period beginning from around the late 1990s saw some recovery of the agricultural sectors in most EU transition countries, with the consequent increase in production and farm input sales, including pesticides.

The growth in pesticide sales for some countries over the past decade has been mainly driven by increasing crop production, but especially the horticulture and vine sub-sectors, for example, this in part explains recent increases in pesticide sales for **Chile**, **Estonia**, **Finland**, **Hungary**, **Iceland**, **Mexico**, **New Zealand**, **Spain** and **Turkey** (Figures 5.1 and 5.2). For some other countries, for example **Finland**, the switch to agri-environmental payments requiring adoption of environmental farm management practices, such as conservation tillage which usually correlates with greater sales of herbicides, has been a major influence in increasing pesticide sales. This development, however, has to be viewed in terms of other environmental benefits from conservation tillage, such as lowering soil erosion rates.

There is evidence that for a growing number of countries, ***the growth in crop production has been decoupled from the sales of pesticides.*** In other words crop production has been increasing at a faster rate over the period since 2000 than the change in pesticide sales (Figure 5.3 Panel A and Panel B). This development, which suggests improvements in the efficiency of pesticide sales per tonne of crop output, was already evident for some countries over the 1990s, but has become more widespread for other countries over the past decade, such as **Australia**, **Belgium**, **Canada**, **France**, **Italy**, **Japan**, **Korea**, **Mexico**, **Netherlands**, **Portugal**, **Slovenia**, the **United Kingdom** and the **United States** (Figure 5.3 Panel B). The correlation between crop production and pesticide sales, however, needs to be treated cautiously as, for example, it does not take into account the toxicity of pesticides, and site specific conditions, such as soil, weather and pest pressures.

The apparent improvements in pesticide sales efficiency for a growing number of OECD countries can be explained by a combination of factors which vary in importance between countries. The main factors include: farmer education and training; the overall decoupling of support from production and input related support (Figure 2.2, Panel A and Panel B; and Figure 2.3); the use of payments to encourage adoption of beneficial pest management practices; pesticide taxes; the use of new pesticide products in lower doses and more targeted; and the expansion in organic farming (Figure 3.7) (OECD, 2012; OECD, 2008). Some countries made earlier progress in the 1990s in adopting these measures. They made such improvements as adopting new pesticide products or pest management practices to reduce the sales of pesticides at a faster rate than the change in crop production, notably (e.g. **Austria**,

Figure 5.1. **Pesticide sales, OECD countries, 1990-2010**

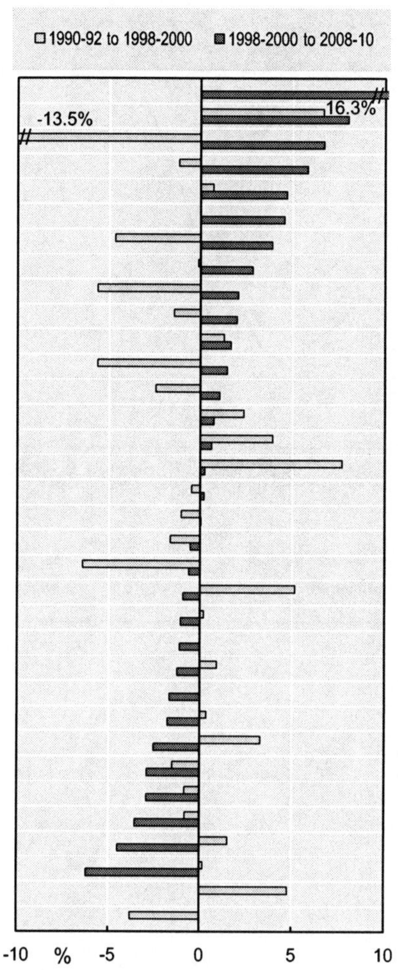

□ 1990-92 to 1998-2000 ■ 1998-2000 to 2008-10

-13.5%

16.3%

-10 % -5 0 5 10

	Average (tonnes of active ingredients)			Average annual % change	
	1990-92[1]	1998-2000[2]	2008-10[3]	1990-92 to 1998-2000	1998-2000 to 2008-10
Chile	..	26 833	57 058	..	16.3
Estonia	164	227	491	6.7	8.0
Hungary	18 554	5 832	11 176	-13.5	6.7
New Zealand	3 490	3 182	5 294	-1.2	5.8
Turkey	11 967	12 550	18 130	0.7	4.7
Iceland	..	4	5	..	4.6
Finland	1 688	1 150	1 688	-4.7	3.9
Spain	36 849	36 476	43 154	-0.1	2.8
Czech Republic	6 699	4 212	5 166	-5.6	2.1
Slovak Republic	3 694	3 330	4 059	-1.5	2.0
Ireland	2 035	2 256	2 661	1.3	1.7
Denmark	4 948	3 111	3 602	-5.6	1.5
Austria	4 206	3 440	3 824	-2.5	1.1
Mexico	31 551	38 037	39 741	2.4	0.7
Greece	8 337	10 921	11 332	3.9	0.6
Australia	19 323	34 963	35 496	7.7	0.3
Germany	32 618	31 402	32 084	-0.5	0.2
Norway	772	709	713	-1.1	0.1
Sweden	1 897	1 660	1 572	-1.7	-0.5
Netherlands	17 354	10 191	9 568	-6.4	-0.6
Portugal	13 190	16 132	14 947	5.2	-0.9
OECD	**908 959**	**923 075**	**826 688**	**0.2**	**-1.1**
United States	325 226	325 377	300 429	0.0	-1.1
Italy	79 843	83 629	75 483	0.9	-1.3
Slovenia	..	1 344	1 172	..	-1.7
EU15	**339 207**	**348 739**	**291 032**	**0.3**	**-1.8**
Canada	32 775	42 500	36 358	3.3	-2.6
Japan	89 112	78 741	60 291	-1.5	-2.9
Korea	26 425	24 676	18 312	-0.9	-2.9
Belgium	10 204	9 538	6 648	-0.8	-3.5
France	95 281	107 649	68 053	1.5	-4.5
United Kingdom	30 754	31 182	16 418	0.2	-6.2
Poland[4]	6 254	8 672	19 520	4.8	n.c.
Switzerland[5]	2 120	1 557	2 188	-3.8	n.c.

..: not available; n.c.: not calculated.

Notes: Countries ranked in descending order according to average annual percentage change 1998-2000 to 2008-10.

For all countries, the data represent pesticide sales, except for the following countries: Korea and Mexico (national production data) and United Kingdom (amount of active substance applied on-farm).

The data are expressed in tonnes of active ingredients except for Chile for which data are expressed in tonnes of formulated product.

Pesticide sales cover agriculture and non-agricultural sales (e.g. forestry, gardens), except for Finland which does not include forestry and for the following countries which only include agriculture: Belgium, Denmark and Sweden.

The following countries are not included in the figure: Israel (time series incomplete), Luxembourg (included in Belgium).

The OECD total does not include: Chile, Iceland, Israel, Poland, Slovenia and Switzerland.

For Israel, sales were 6 946 tonnes for 2008-10.

1. Data for 1990-92 average equal the 1991-93 average for Greece, Norway, Poland and Slovak Republic; the 1993-95 average for Estonia and Turkey; the 1995-97 average for Italy; and the 1996-98 average for Portugal.
2. Data for 1998-2000 average refer to the 1999-2001 average for Iceland; the 2000-02 average for Italy, Portugal, Slovenia and Turkey; and the 2001-03 average for Chile.
3. Data for 2008-10 average refer to the 2004-06 average for Australia, Canada, Greece, Mexico and Spain; the 2005-07 average for the United States; the 2006-08 average for Chile and Iceland; and the 2007-09 average for Japan and New Zealand.
4. Break in time series from 2005, data not comparable.
5. Break in time series from 2006, data not comparable.

Source: OECD Environmental Compendium Database 2008, www.oecd.org/environment; Statistical Office of the European Community (EUROSTAT), *http://epp.eurostat.ec.europa.eu*; and national data.

StatLink http://dx.doi.org/10.1787/888932792768

Figure 5.2. **Pesticide sales index, OECD countries, 1990-2010**

Index 1990-92 = 100

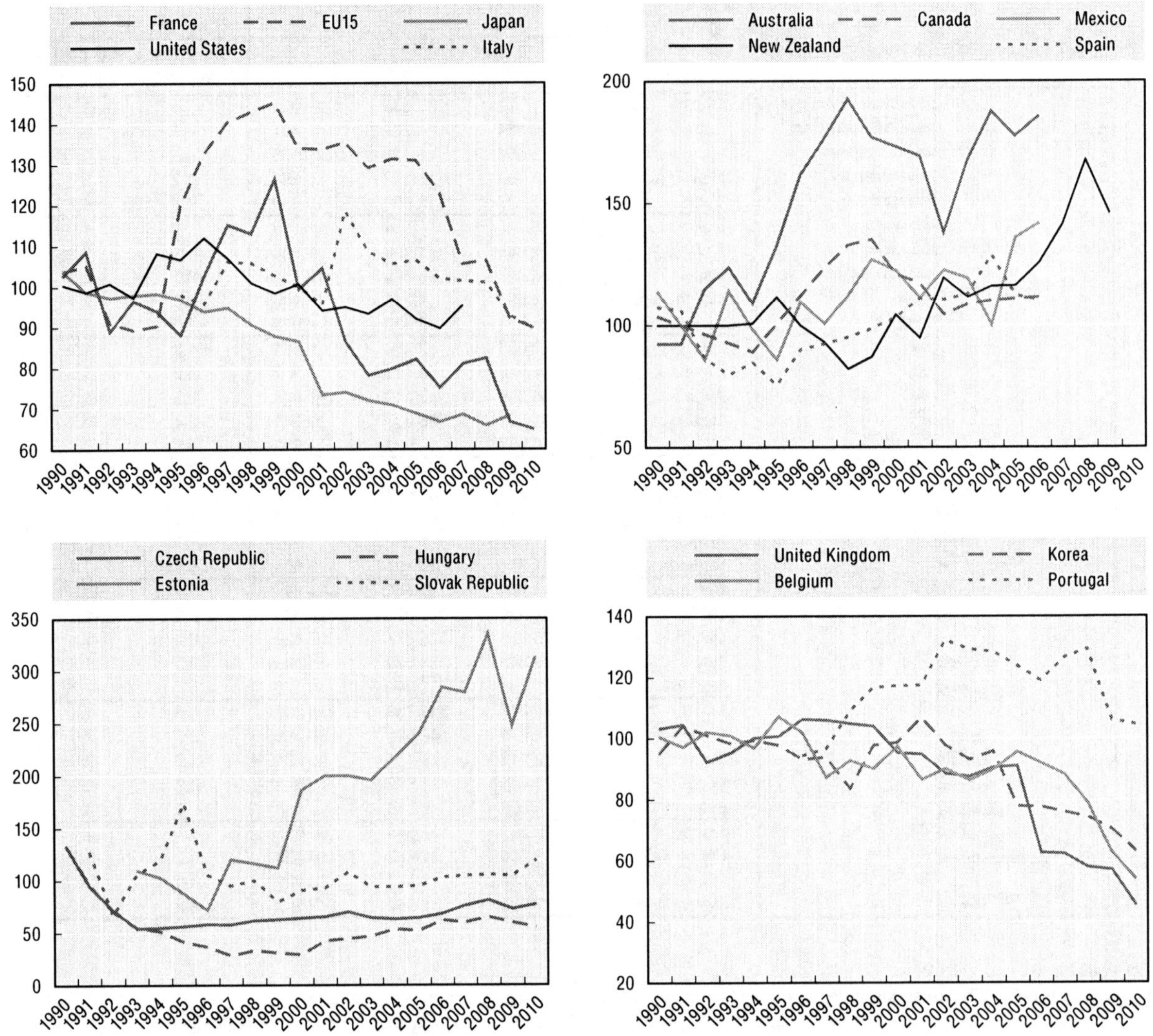

Notes: Caution is required in comparing trends across countries because of differences in data definitions, coverage and time periods.

The Index 1990-92 = 100 equals the 1991-93 = 100 for Slovak Republic; the 1993-95 = 100 for Estonia; the 1996-98 = 100 for Portugal; and the 1995-97 = 100 for Italy.

For all countries, the data represent pesticide sales, except for the following countries: Korea and Mexico (national production data) and United Kingdom (amount of active substance applied on-farm).

Pesticide sales cover agriculture and non-agricultural sales, except for Belgium which only includes agriculture.

Source: OECD Environmental Compendium Database 2008, www.oecd.org/environment; Statistical Office of the European Community (EUROSTAT), *http://epp.eurostat.ec.europa.eu*; and national data.

StatLink *http://dx.doi.org/10.1787/888932792787*

Denmark, **Finland**, **Germany**, **Hungary**, **Korea**, **Netherlands**, **New Zealand**, **Norway**, **Slovak Republic**, **Spain**, **Sweden**, **Switzerland**, **Turkey** and the **United States**) (Figure 5.3).

There are no comparable cross country data on the risks to human health and the environment from the use of pesticides in agriculture, although there is a considerable body of research in the area, and a few countries have developed their own pesticide risk indicators (OECD, 2008). This lack of information on pesticide risks is further compounded in terms of a lack of comprehensive knowledge and information on the health and environmental effects

Figure 5.3. **Pesticide sales and crop production volume, OECD countries, 1990-2010**

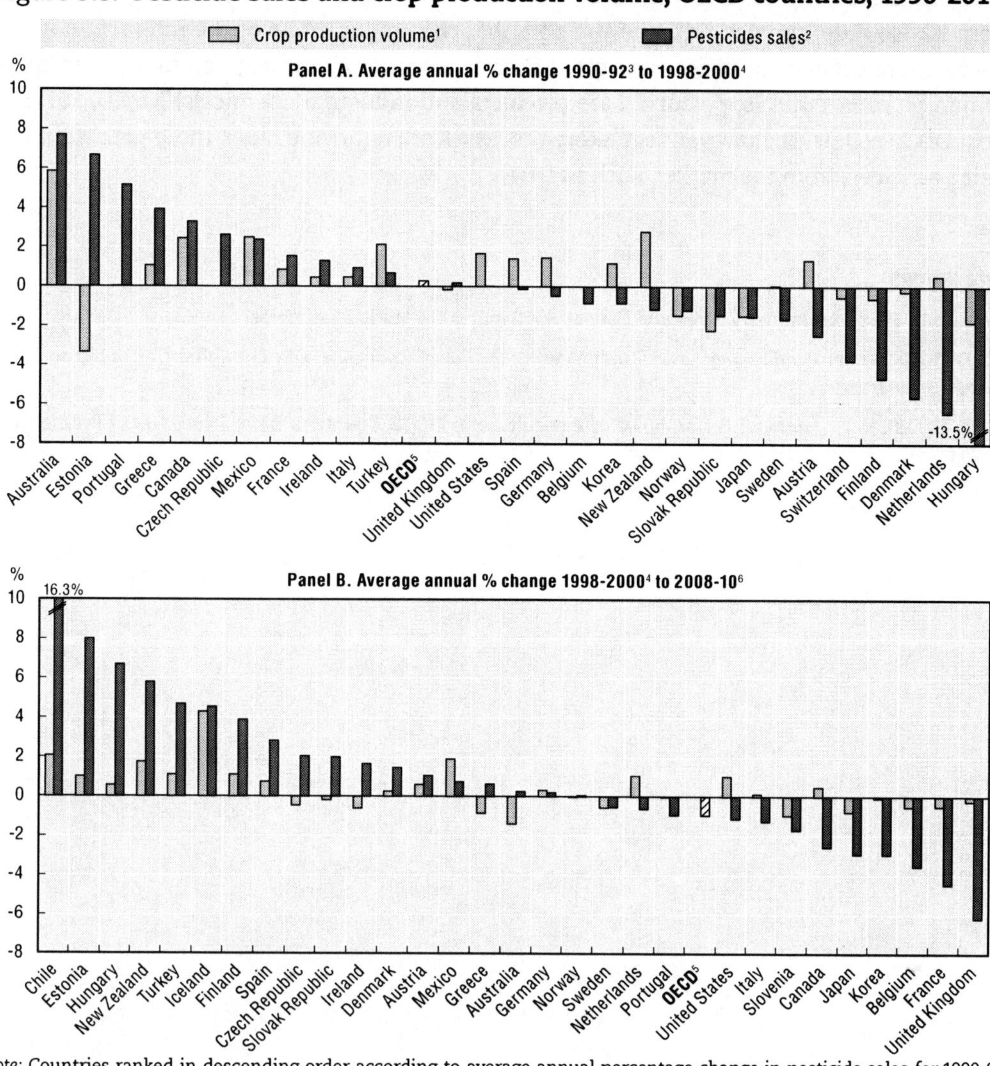

Note: Countries ranked in descending order according to average annual percentage change in pesticide sales for 1990-92 to 1998-2000 and 1998-2000 to 2008-10, respectively.

1. The FAO indices of crop production show the relative level of the aggregate volume of crop production for each year in comparison with the base period 2004-06. They are based on the sum of price weighted quantities of different crop commodities produced after deductions of quantities used as seed and feed weighted in a similar manner. The resulting aggregate represents, therefore, disposable production for any use except as seed and feed. All the indices at the country, regional and world levels are calculated by the Laspeyres formula. Production quantities of each commodity are weighted by 2004-06 average international commodity prices and summed for each year. To obtain the index, the aggregate for a given year is divided by the average aggregate for the base period 2004-06. Due to technical reasons it is not possible to provide an OECD or EU average.
2. Pesticide sales cover agriculture and non-agricultural uses (e.g. forestry, gardens), except for Finland which does not include forestry and for the following countries which only include agriculture: Belgium, Denmark and Sweden. For all countries, the data represent pesticide sales, except for the following countries: Korea and Mexico (national production data) and United Kingdom (amount of active substance applied on-farm). The data are expressed in tonnes of active ingredients except for Chile, for which data are expressed in tonnes of formulated product. The following countries are not included in the figure: Israel (time series are incomplete), Luxembourg (included in Belgium), Poland (break in time series from 2005, data not comparable), and Switzerland (break in time series from 2006, data not comparable).
3. Crop production and pesticides data for 1990-92 average equal the 1991-93 average for Greece, Norway, Poland and Slovak Republic; the 1993-95 average for Estonia and Turkey; the 1995-97 average for Italy; and the 1996-98 average for Portugal.
4. Pesticide sales data for 1998-2000 average refer to the 1999-2001 average for Iceland; the 2000-02 average for Italy, Portugal, Slovenia and Turkey; the 2001-03 average for Chile; and the year 2000 for Belgium for crop production data.
5. The OECD total for pesticides sales does not include Chile, Iceland, Israel, Slovenia and Switzerland.
6. Pesticide sales data for 2008-10 average refer to the 2004-06 average for Australia, Canada, Greece, Mexico and Spain; the 2005-07 average for United States; the 2006-08 average for Chile and Iceland; and the 2007-09 average for Japan and New Zealand.

Source: FAOSTAT (2012), *http://faostat.fao.org*; *OECD Environmental Compendium Database 2008*, *www.oecd.org/environment*; Statistical Office of the European Community (EUROSTAT), *http://epp.eurostat.ec.europa.eu*; and national data.

StatLink http://dx.doi.org/10.1787/888932792806

with the release of mixtures of pesticides rather than a single pesticide product. Moreover, there is also little understanding of the potential risk implications of the interaction in the environment between pesticides and other chemical contaminants (e.g. veterinary medicines, human pharmaceuticals, personal care products and industrial chemicals) (OECD, 2012). In most OECD countries, however, regulatory processes are removing older, more persistent and toxic pesticides, from the market, such as DDT.

References

EPA (2011), *Pesticides Industry Sales and Usages 2006 and 2007 Market Estimates.*

OECD (2012), *Water Quality and Agriculture: Meeting the Policy Challenge*, OECD Publishing, *www.oecd.org/agriculture/water.*

OECD (2008), *Environmental Performance of Agriculture in OECD Countries Since 1990*, OECD Publishing, France.

Chapter 6

Energy: On-farm energy consumption and production of biofuels from agricultural feedstocks

This chapter reviews the environmental performances of agriculture in OECD countries related to on-farm energy consumption and production of biofuels from agricultural feedstocks. It provides a description of the policy context (issues and main challenges), definitions for the agri-environmental indicators presented, and elements related to concepts, interpretations, links to other indicators, as well as measurability and data quality. The chapter then describes the main trends of the agri-environmental indicators, using available data covering the period 1990-2010 and based on a set of tables and figures.

6.1. Policy context

The issue

Agriculture can play a double role in relation to energy, both as a consumer and producer of energy. Farming is a direct energy consumer for crop and livestock production, and also consumes energy indirectly in terms of the energy required to produce fertilisers, pesticides, machinery and other inputs. But agriculture can also produce energy and raw materials through biomass production as a feedstock to generate bioenergy, including biofuels, mainly bioethanol and biodiesel (OECD, 2008). Agriculture can also provide land on which energy can be generated, such as from wind and solar sources.

Support to agricultural energy use is widespread across OECD countries. This typically involves reducing the standard rate of fuel tax for on-farm consumption, but also for power and heat in some cases. Support is also common across OECD countries for bioenergy, by providing agriculture a combination of tax incentives and payments for the production of bioenergy feedstocks using agricultural raw materials (e.g. maize) and waste (e.g. straw) (OECD, 2011a; OECD, 2011b).

Support that reduces on-farm fuel costs may act as a disincentive to reduce on-farm energy consumption and use energy more efficiently, and also, by stimulating higher energy use, put pressure on the environment by leading to increased greenhouse gas emissions and other air pollutants from agriculture. Similarly biofuel support policies, in particular where these are associated with their production from cereals, sugar and oilseed crops, are not the most efficient way of addressing a number of environmental objectives, such as reducing fossil fuel use, water pollution and greenhouse gas emissions (OECD, 2011a; OECD, 2012).

Main challenges

The key challenge for agriculture with regard to energy is to improve energy use efficiency on-farm through lowering energy consumption per unit of agricultural production, and also seeking opportunities to increase production of biofuel feedstocks that are environmentally neutral (i.e. requires less energy to produce than the energy generated and has minimal impact in terms of water pollution, air pollution, etc., in producing the feedstock).

There are a broader set of challenges regarding energy in relation to agriculture, including the energy consumption along the agro-food chain, both to provide energy for inputs used by agriculture (e.g. fertilisers, pesticides, machinery), and in processing, transporting and marketing agricultural commodities (e.g. food, feed, fibre). Similarly biofuel production from agricultural feedstocks raises a number of concerns, for example, the competition for land to produce food, feed, fibre and feedstocks for energy production. These broader considerations, however, are not the focus of the indicators in this chapter which only relate to primary agriculture, and not the agro-food chain or other broader considerations.

6.2. Indicators

Definitions

The indicators related to agricultural energy consumption and production include changes in:

- Direct on-farm energy consumption.
- Biofuel production, to produce bioethanol and biodiesel from agricultural feedstocks.

Concepts, interpretation, limitations and links to other indicators

Purchased energy is essential to provide power for modern agricultural production systems, as it is for most other industrial sectors. From an environmental perspective, however, agricultural energy consumption, as with other fossil fuel energy using sectors, can lead to air pollution through emission of greenhouse gases (principally carbon dioxide, CO_2); emissions of nitrogen oxides (NO_X), sulphur dioxide (SO_2), particulate matter; as well as emissions of ozone depleting precursors. While energy produced from fossil fuel combustion is non-renewable, renewable energy derived from agricultural biomass feedstock has the potential to provide environmental benefits, for example, some feedstocks are carbon neutral from a climate change perspective.

The major limitation of the on-farm energy consumption indicator for most countries, concerns the difficulty of separating agricultural energy consumption data from data for energy consumption by hunting, forestry and fisheries. Also, the extent to which farm household consumption is included in the data is unknown. Therefore, caution is required when comparing agricultural energy consumption trends across countries, especially where forestry, fisheries or hunting may be significant activities.

In the case of the bioethanol and biodiesel, an important limitation is the assumption that the feedstocks used to produce these fuels all originate from primary agriculture, when some other feedstock sources might be used to produce these fuels, such as farm and forestry by-products and waste. In future agricultural energy production indicators might be extended to include solar and wind sources where these are located on agricultural land.

Direct on-farm energy consumption acts primarily as a driving force on the state of climate change through greenhouse gas emissions (Chapter 11), although emissions of CO_2 from fossil fuel energy use is a minor contributor of agricultural greenhouse gas emissions compared to methane and nitrous oxide (Table 11.1). There are also secondary environmental concerns with regard to energy consumption in agriculture, related to air pollution from burning fossil fuels, such as particulate matter and ozone depletion. The production of biofuel feedstocks can have a wide spectrum of impacts on the environment (e.g. water and air pollution, biodiversity), depending on the farm practices and systems used in their production, as discussed later in this chapter.

Measurability and data quality

The OECD agricultural energy indicator in this section focuses on *direct* on-farm energy consumption by primary agriculture, which includes energy consumption for irrigation, drying, horticulture, machinery and livestock housing. The data and definition of on-farm energy consumption are drawn from Eurostat and the International Energy Agency, while data for biodiesel and bioethanol production is largely drawn from national sources and the *OECD-FAO Agricultural Outlook Database*.

The main concern with data quality for direct on-farm energy consumption is that it includes for most countries, the hunting, forestry and fisheries sectors, and is not available for primary agriculture. In this regard interpretation of the trends discussed here need to be treated with caution. Moreover, some national biofuel production may rely heavily on imported feedstocks rather than domestically produced feedstocks. In **Portugal**, for example, over 90% of biofuel production is based on imported raw materials.

6.3. Main trends

Direct on-farm energy consumption declined over the period 2000 to 2010 compared to an increasing trend over the 1990s (Figure 6.1). To some extent this reflects the slowdown in OECD agricultural production over the same period (Figure 3.1). There are, however, large disparities in terms of trends in direct on-farm energy consumption over the most recent decade across OECD countries, with the **EU15**, **Japan**, **Korea** and **Poland** accounting for much of the decrease in OECD consumption, in part, explained by the lower growth in the agricultural sectors for these countries over the 2000s and also improvements in overall energy use efficiency on-farm.

Significant increases in direct on-farm energy consumption, however, are evident in, for example, **Australia, Israel, Mexico, New Zealand, Spain, Turkey** and the **United States** over the 10 years from 2000 (Figure 6.1). The growth in consumption for these countries, can be related to a combination of: rising agricultural production over the 2000s, in most cases; the continuing expansion of mechanisation and in machinery power; and substituting labour for machinery. The relative importance of these different factors, however, varied between these countries.

The share of primary agriculture in total national energy consumption is extremely low for most OECD countries between 0.4-6.3% (2008-10), with only six countries having a share of 4-6% (Figure 6.1). But despite the low share in national energy consumption, the agricultural sector in OECD countries is vulnerable to changes in crude oil prices, given the sector's reliance on energy and the energy embodied in the inputs used in agricultural production. Energy support provided to farmers in some countries to an extent, however, shields farmers from the variability of world oil prices, and may also discourage greater energy efficiency gains (OECD, 2008).

Improving energy efficiency in primary agriculture is, however, taking on an increasingly important role for nearly all countries, not only in terms of the need to reduce overall farm operating costs, but also as part of national programmes to lower greenhouse gas emissions from the use of fossil fuels. Cross country data on energy efficiency changes in primary agriculture are not possible with the current OECD dataset, as for most countries the data also includes, hunting, forestry and fisheries. Limited research evidence, however, suggests energy efficiency in primary agriculture over the past two decades has improved or remained stable (see for example **Switzerland**, Chapter 13).

OECD biofuel production, largely from agricultural feedstocks, is dominated by the **United States** (mainly bioethanol, but also biodiesel), and to a lesser extent **Australia, Canada, France, Germany, Italy** and **Poland** (Figures 6.2 and 6.3, Tables 6.1 and 6.2). While production of biofuels has a long history in some countries, for most countries production has expanded rapidly over the period from 2000 to 2010 (Tables 6.1 and 6.2). Bioethanol production dominants biofuel production in OECD countries, accounting for 77% of total OECD biofuel production in 2008-10, converted in energy terms (kilo tonne of oil equivalent, ktoe).

Figure 6.1. **Direct on-farm energy consumption, OECD countries, 1990-2010**

□ 1990-92 to 1998-2000
■ 1998-2000 to 2008-10
-21.8%
-10.2%
-10 % 0 10

	Average			Average annual % change		Share of agriculture in total national energy consumption	Share of country in OECD total agricultural energy consumption
	1990-92	1998-2000	2008-10	1990-92 to 1998-2000	1998-2000 to 2008-10	2008-10	2008-10
	Thousand tonnes oil equivalent			%		%	%
Luxembourg	12	12	28	0.0	9.0	0.7	0.0
Turkey	1 975	2 941	5 114	5.1	5.7	6.3	8.6
Estonia	475	67	94	-21.8	3.5	3.2	0.2
Australia	1 320	1 576	2 136	2.2	3.1	2.8	3.6
Mexico	2 248	2 766	3 517	2.6	2.4	3.1	5.9
New Zealand	362	465	582	3.2	2.3	4.6	1.0
Israel	87	134	148	5.5	1.0	1.0	0.3
Spain	1 796	2 236	2 464	2.8	1.0	2.7	4.2
United States	14 819	13 972	15 319	-0.7	0.9	1.0	25.9
Finland	856	735	803	-1.9	0.9	3.2	1.4
Slovenia[1]	n.a.	77	79	n.a.	0.3	1.6	0.1
Italy	2 801	2 957	3 006	0.7	0.2	2.4	5.1
Austria	581	541	550	-0.9	0.2	2.0	0.9
Belgium	582	841	828	4.7	-0.2	2.3	1.4
OECD[2]	**57 862**	**60 894**	**59 184**	**0.6**	**-0.3**	**1.6**	**100**
France	3 515	3 788	3 658	0.9	-0.3	2.3	6.2
Denmark	759	746	714	-0.2	-0.4	4.7	1.2
Canada	3 376	4 132	3 927	2.6	-0.5	2.0	6.6
Ireland	257	301	282	2.0	-0.7	2.3	0.5
EU15	**21 673**	**21 253**	**19 374**	**-0.2**	**-0.9**	**2.0**	**33**
Switzerland	193	312	270	6.2	-1.4	1.3	0.5
Sweden	812	781	668	-0.5	-1.6	2.0	1.1
Netherlands	3 574	3 830	3 274	0.9	-1.6	6.3	5.5
Czech Republic	1 393	619	527	-9.6	-1.6	2.1	0.9
Greece	1 060	1 090	921	0.4	-1.7	4.5	1.6
Poland	3 635	4 719	3 641	3.3	-2.6	5.8	6.1
United Kingdom	1 297	1 215	902	-0.8	-2.9	0.7	1.5
Hungary	947	698	488	-3.7	-3.5	2.9	0.8
Germany	3 311	1 524	926	-9.2	-4.9	0.4	1.6
Japan	2 307	2 909	1 727	2.9	-5.1	0.5	2.9
Slovak Republic	544	231	135	-10.2	-5.2	1.2	0.2
Korea	1 852	3 381	1 808	7.8	-6.1	1.2	3.1
Portugal	461	655	350	4.5	-6.1	1.9	0.6
Norway	588	640	334	1.1	-6.3	1.8	0.6
Iceland	68	80	42	2.1	-6.3	1.4	0.1

Notes: Countries are ranked in descending order according to average annual percentage change 1998-2000 to 2008-10.
Data cover total on-farm energy consumption by primary agriculture (for irrigation, drying, horticulture, machinery and livestock housing), forestry and hunting.
Data for Austria, Mexico, and Turkey include agriculture and fisheries. Data for Australia include agriculture only.
The statistical data for Israel are supplied by and under the responsibility of the relevant Israeli authorities. The use of such data by the OECD is without prejudice to the status of the Golan Heights, East Jerusalem and Israeli settlements in the West Bank under the terms of international law.
1. For Slovenia, 1990-99 data are not available; data for the 1998-2000 average refer to 2000-02.
2. OECD total excludes Chile and Slovenia.
Source: IEA (2012), International Energy Agency Data Services, *http://dotstat.oecd.org/Index.aspx* (energy > world energy balances); EUROSTAT, Statistical Office of the European Community, *http://epp.eurostat.ec.europa.eu*; and national sources for New Zealand, Turkey and United Kingdom.

StatLink http://dx.doi.org/10.1787/888932792825

The feedstocks to produce biofuels in OECD countries are largely maize in the **United States** to produce bioethanol, and in the **European Union** rapeseed oil is mainly used to produce biodiesel. The composition of agricultural feedstocks to produce biofuels in the coming decade could diversify with greater use of wheat, sugar beet, and other

Figure 6.2. **Bioethanol production, OECD countries, 2008-10**

% country share in OECD total

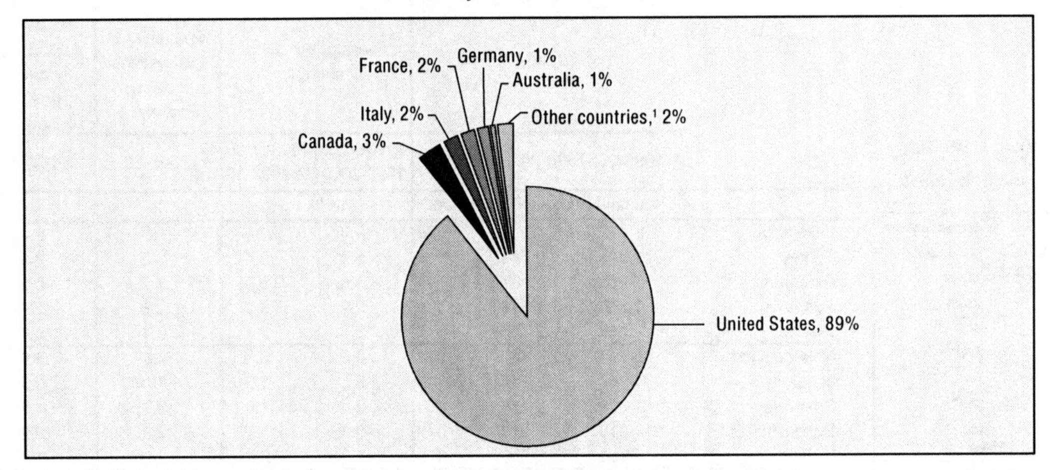

1. Other countries include Austria, Czech Republic, Hungary, Japan, Korea, Mexico, New Zealand, Poland and Turkey.

Source: OECD-FAO Agricultural Outlook 2012-2021, www.agri-outlook.org; and national sources for Austria, Czech Republic, France, Germany, Hungary, Italy, Japan, New Zealand and Poland.

StatLink http://dx.doi.org/10.1787/888932792844

Figure 6.3. **Biodiesel production, OECD countries, 2008-10**

% country share in OECD total

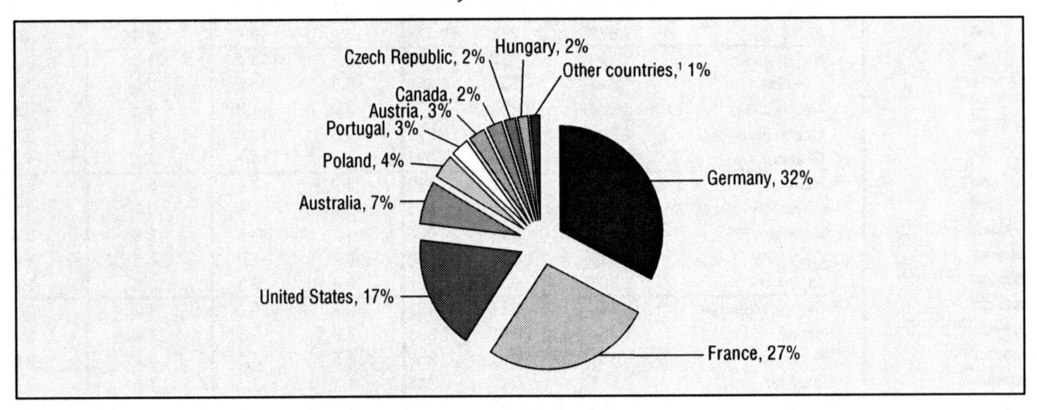

1. Other countries include Italy, New Zealand, Slovenia and Turkey.

Source: OECD-FAO Agricultural Outlook 2012-2021, www.agri-outlook.org; and national sources for Austria, Czech Republic, France, Germany, Hungary, Italy, New Zealand, Poland, Portugal and Slovenia.

StatLink http://dx.doi.org/10.1787/888932792863

vegetable oils, and possibly growth in the use of so called "second generation" biofuels from cellulosic farm by-products and waste, such as cereal straw (OECD, 2011b).

The expansion of biofuel production over the past decade in OECD countries has been mainly driven by policy support in the form of mandates or targets that impact use; tax relief for producers and consumers of biofuels; border protection measures; fuel quality specifications; and support for investment production capacity (OECD, 2011b). Both bioethanol production in the **United States** and biodiesel production in the **European Union** are projected to continue to expand in the decade to 2020, driven by similar policy support provided over the past decade (OECD, 2011b).

A key conclusion from most studies on the links between biofuel production from agricultural feedstocks on the environment (mainly soils, water systems, air emissions and

Table 6.1. **Bioethanol production, OECD countries, 1990-2011**

Thousand tonnes oil equivalent (toe)

	1990	1991	1992	1993	1994	1995	1996	1997	1998	1999	2000	2001	2002	2003	2004	2005	2006	2007	2008	2009	2010	2011
Australia	-	-	-	-	-	-	-	-	-	-	20	20	20	26	66	66	82	92	97	157	204	235
Austria	-	-	-	-	-	-	-	-	-	-	-	-	-	-	-	-	-	11	62	122	138	151
Canada	-	-	-	-	-	-	20	20	90	96	195	190	189	190	193	226	379	470	700	748	820	903
Czech Republic					-	-	-	-	-	-	-	-	-	-	-	-	1	17	39	58	61	-
France	-	-	-	18	25	24	39	54	63	58	59	58	58	50	52	75	149	274	378	460	478	429
Germany	-	-	-	-	-	-	-	-	-	-	-	-	-	-	13	84	219	200	297	382	376	-
Hungary	-	-	-	-	-	-	-	-	-	-	-	-	-	-	-	-	-	28	47	50	58	-
Italy	-	-	-	-	-	-	-	-	-	-	-	-	-	-	-	-	-	-	429	509	468	-
Japan	-	-	-	-	-	-	-	-	-	-	-	-	56	70	58	58	58	56	56	51	54	-
Korea	-	-	-	-	-	-	-	-	-	-	-	105	88	89	84	84	83	89	82	86	88	88
Mexico	-	-	-	-	22	22	22	22	22	23	28	29	24	20	18	30	25	31	31	31	36	38
New Zealand	-	-	-	-	-	-	-	-	-	-	-	-	-	-	-	-	-	0	1	2	2	2
Poland	-	-	-	-	14	32	51	56	51	45	26	35	42	39	25	57	83	61	56	84	104	-
Turkey	-	-	-	-	-	-	-	-	-	-	-	-	-	-	10	24	26	23	28	33	38	39
United States	-	-	-	-	-	-	2 124	2 510	2 678	2 842	3 263	3 722	4 683	6 241	7 130	8 606	10 966	15 272	18 231	22 621	24 720	26 082
OECD	-	-	-	**18**	**60**	**78**	**2 255**	**2 662**	**2 904**	**3 064**	**3 592**	**4 160**	**5 161**	**6 725**	**7 649**	**9 309**	**12 070**	**16 622**	**20 533**	**25 393**	**27 645**	**n.c.**
EU15	-	-	-	**18**	**25**	**24**	**39**	**54**	**63**	**58**	**59**	**58**	**58**	**50**	**65**	**159**	**368**	**484**	**1 166**	**1 472**	**1 461**	**n.c.**

n.c.: not calculated.

Notes: Production less than one thousand toe or no production for Belgium, Chile, Czech Republic (from 1990 to 2005), Denmark, Estonia, Finland, Iceland, Ireland, Israel, Luxembourg, Norway, Portugal, Slovak Republic, Slovenia, Spain, Sweden, Switzerland, United Kingdom, and Turkey (from 1990 to 2003).

Data converted from litres to toe using a conversion rate 1 000 litres of bioethanol = 0.51 toe, drawn from Eurostat, see *http://epp.eurostat.ec.europa.eu*.

Source: *OECD-FAO Agricultural Outlook 2012-2021*, *www.agri-outlook.org*; national sources for Austria, Czech Republic, France, Germany, Hungary, Italy, Japan, New Zealand and Poland.

StatLink http://dx.doi.org/10.1787/888932793452

Table 6.2. **Biodiesel production, OECD countries, 1990-2011**

Thousand tonnes oil equivalent (toe)

	1990	1991	1992	1993	1994	1995	1996	1997	1998	1999	2000	2001	2002	2003	2004	2005	2006	2007	2008	2009	2010	2011
Australia	-	-	-	-	-	-	-	-	-	-	-	-	-	-	-	38	68	236	473	495	500	506
Austria	-	-	-	-	-	-	-	-	-	-	-	-	-	-	-	-	78	154	159	206	215	198
Canada	-	-	-	-	-	-	-	-	-	-	-	-	-	2	3	21	41	76	110	198	246	259
Czech Republic	-	-	3	5	8	10	17	24	14	27	59	62	91	99	75	111	97	72	67	136	174	-
France	-	-	-	8	65	155	219	253	222	242	303	303	329	378	396	433	597	963	1 778	2 107	2 014	1 836
Germany	-	-	-	-	-	-	-	-	-	-	195	201	399	665	869	1 285	2 127	2 562	2 500	2 216	2 482	-
Hungary	-	-	-	-	-	-	-	-	-	-	-	-	-	-	-	-	-	2	118	123	117	-
Italy	-	-	-	-	-	-	-	-	-	-	-	-	-	-	-	-	-	-	33	91	41	-
New Zealand	-	-	-	-	-	-	-	-	-	-	-	-	-	-	-	-	-	1	1	1	1	2
Poland	-	-	-	-	-	-	-	-	-	-	-	-	-	-	-	57	81	39	148	323	328	-
Portugal	-	-	-	-	-	-	-	-	-	-	-	-	-	-	-	-	-	168	130	227	285	-
Slovenia	-	-	-	-	-	-	-	-	-	-	-	-	-	-	-	-	-	3	5	5	12	-
Turkey	-	-	-	-	-	-	-	-	-	-	-	-	-	-	-	399	399	399	22	20	103	25
United States	-	-	-	-	-	-	-	-	-	-	12	22	27	27	92	252	639	1 168	1 849	1 286	743	2 310
OECD	**-**	**-**	**3**	**13**	**73**	**165**	**235**	**278**	**236**	**269**	**569**	**589**	**847**	**1 171**	**1 434**	**2 596**	**4 126**	**5 842**	**7 394**	**7 433**	**7 261**	**n.c.**
EU15	**-**	**-**	**-**	**8**	**65**	**155**	**219**	**253**	**222**	**242**	**498**	**504**	**728**	**1 043**	**1 265**	**1 719**	**2 802**	**3 847**	**4 600**	**4 847**	**5 036**	**n.c.**

n.c.: not calculated.

Notes: Production less than one thousand toe or no production for Belgium, Chile, Denmark, Estonia, Finland, Greece, Iceland, Ireland, Israel, Japan, Luxembourg, Mexico, Netherlands, Norway, Slovak Republic, Spain, Sweden, Switzerland, United Kingdom and Turkey (from 1990 to 2004).

Data converted from litres to toe using a conversion rate 1 000 litres of biodiesel = 0.78 toe, drawn from Eurostat, *http://epp.eurostat.ec.europa.eu.*

Source: OECD-FAO *Agricultural Outlook 2012-2021*, *www.agri-outlook.org*; national sources for Austria, Czech Republic, France, Germany, Hungary, Italy, New Zealand, Poland, Portugal and Slovenia.

StatLink http://dx.doi.org/10.1787/888932793471

biodiversity) is that in general feedstocks from annual crops, such as maize and rapeseed, can have a more damaging impact on the environment than second generation feedstocks, such as reed canary grass and short rotation woodlands.

Another important conclusion is that the location of production and the type of tillage practice, crop rotation system and other farm management practices used in producing feedstocks for biofuel production, will also greatly influence environmental outcomes. Moreover, the increasing production of biofuels from agricultural and food wastes and residues (e.g. straw, manure, food waste, animals fats), may help to lower the demand for production of feedstocks from cultivated crops and hence, reduce environmental impacts. But a note of caution is important here, as the potential impact on the environment from growing agricultural feedstocks for bioenergy production have not been fully evaluated, while the implications of competition for land between bioenergy feedstock and food production are not fully understood (OECD, 2010; 2011b; 2012).

References

OECD (2012), *Water Quality and Agriculture: Meeting the Policy Challenge*, OECD Publishing.

OECD (2011a), *Agricultural Policy Monitoring and Evaluation 2011: OECD Countries and Emerging Economies*, OECD Publishing.

OECD (2011b), *OECD-FAO Agricultural Outlook 2011-2020*, OECD Publishing.

OECD (2010), *Sustainable Management of Water Resources in Agriculture*, OECD Publishing.

OECD (2008), *Environmental Performance of Agriculture in OECD Countries since 1990*, OECD Publishing, *www.oecd.org/tad/sustainable-agriculture/agri-environmentalindicators.htm*.

Chapter 7

Soil: Water and wind erosion

This chapter reviews the environmental performances of agriculture in OECD countries related to soil, in particular water and wind erosion. It provides a description of the policy context (issues and main challenges), definitions for the agri-environmental indicators presented, and elements related to concepts, interpretations, links to other indicators, as well as measurability and data quality. The chapter then describes the main trends of the agri-environmental indicators, using available data covering the period 1990-2010 and based on a set of tables and figures.

7.1. Policy context

The issue

Soil erosion, mainly through water and wind processes, is one of the most widespread forms of soil degradation across OECD countries. Most OECD member countries have developed programmes promoting practices specifically targeted at reducing the risk of soil erosion, including transfers of arable land to grassland, extensive use of pastures, green cover (mainly in the winter period), and promoting conservation tillage practices. The **European Union** and the **United States** also use programmes promoting the long-term retirement of vulnerable land from agricultural production, while afforestation of agricultural land is promoted in some OECD countries to address soil erosion problems.

Main challenges

The key challenge in addressing soil erosion risks in agriculture is to increase the share of land that is subject to low or tolerable rates of soil erosion to maintain soil health. This is important to avoid threats to the fertility and productivity of agricultural soils from losses of topsoils, and to limit damage to the environment both in terms of soil sediment loss to water or into the air. Limiting accelerated soil loss can also be beneficial toward, for example: retaining the carbon content of soils; lowering risks of flooding and landslides; avoiding the dredging costs of removing soil from rivers, lakes, estuaries, and reservoirs; and reducing the costs of treating water for drinking. In this regard farmers by developing these positive externalities from farming can provide ecosystem services for society more broadly (Figure 1.1).

7.2. Indicators

Definitions

The indicators related to agricultural soil erosion include changes in area of:

- Agricultural land affected by water and wind erosion classified as having moderate to severe water and wind erosion risk.

Concepts, interpretation, limitations and links to other indicators

Soil plays an important role in maintaining a balanced ecosystem and in producing quality agricultural products (OECD, 2003). Agricultural soils provide two key functions: support production (notably agriculture but also forestry, etc.); and environmental functions, such as water filtration and conservation, carbon sequestration, and as a reservoir for biodiversity. There can be a significant time delay between recognising soil degradation and developing conservation strategies, in order to maintain soil health and crop productivity. The intensity of rainfall, degree of protective crop cover, slope and soil type are the controlling factors of water erosion.

Figure 7.1. **Agricultural land area classified as having moderate to severe water erosion risk, OECD countries, 1990-2010**

Risk of water erosion greater than 11 tonnes/hectare/year of soil loss as a percentage share of total agricultural land area

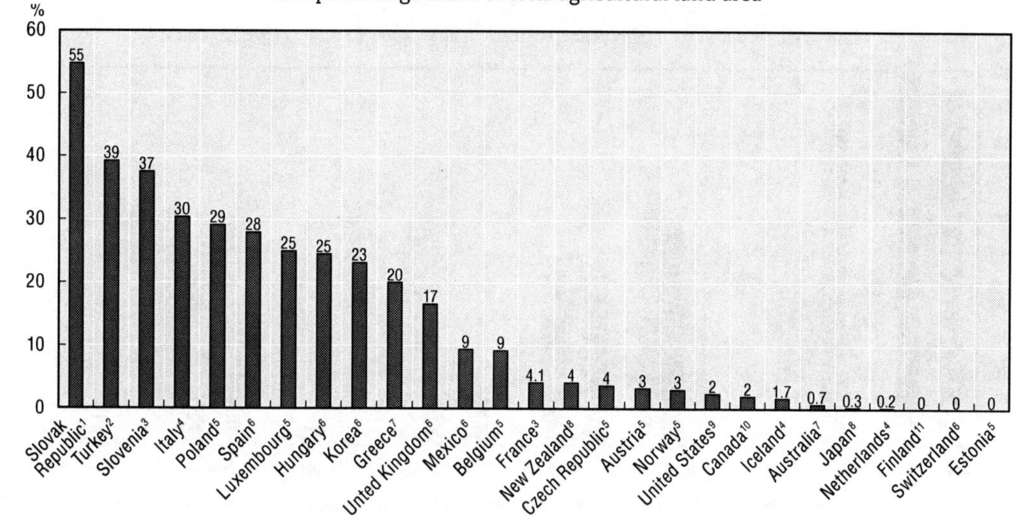

Note: Countries are ranked in terms of highest to lowest % share of agricultural land at risk to water erosion.

1. Data for Slovak Republic refer to 2003-04.
2. Data for Turkey refer to 1990-94.
3. Data for France and Slovenia refer to 2006-07.
4. Data for Iceland, Italy, and Netherlands refer to 1995-99.
5. Data for Austria, Belgium, Czech Republic, Estonia, Luxembourg, Norway and Poland refer to 2009-10.
6. Data for Hungary, Mexico, Spain, Switzerland and United Kingdom refer to 2000-02 and Korea refer to 2002. Soil erosion data for Spain includes agriculture and forestry land. For Switzerland, the total agricultural area includes summer pastures (alpine pastures). For Mexico, the area of risk is the sum of moderate + severe + extreme erosion categories.
7. Data for Australia and Greece drawn from OECD (2008). Data for Greece covers all land, including agricultural land.
8. Data for Japan and New Zealand refer to 1985-89.
9. Data for United States refer to 2007-08.
10. Data for Canada refer to 2005-06, values for cultivated cropland.
11. Data for Finland refer to 2001.

Source: OECD (2008), *Environmental Performance of Agriculture in OECD Countries Since 1990*, *www.oecd.org/agriculture*; Joint Research Centre, European Union; Unpublished Estimates Pan-European RUSLE Model JRC, 2011; and national sources.

StatLink http://dx.doi.org/10.1787/888932792882

The process of wind erosion is also controlled by climate (soil moisture conditions), crop cover and soil type and involves detaching and transporting soil particles (mainly silt and fine sand) over varying distances. Wind erosion is most prevalent in arid and semi-arid areas or where soils can exist in a very dry state for extended periods. Loss of topsoil by erosion also contributes to the loss of nutrients.

Soil tillage practices can also contribute to erosion by moving soil on hilly landscapes, i.e. removing soil from the slopes' top to the bottom (Eilers et al., 2010). Other soil degradation processes, including compaction, acidification, toxic contamination and salinisation largely relate to specific regions in some countries and therefore it is not possible to provide an overview of OECD trends of these soil degradation processes.

Indicators for soil erosion from water are generated by models, most often variants of the Universal Soil Loss Equation (USLE). Although these models take account of soil type, topography, climate and crop cover, they are using generalised inputs that provide estimates of soil erosion risk rather than actual field measurement values. It is important to stress that the trends reported in this chapter only concern on-farm soil erosion.

While the USLE is commonly used by most OECD countries, the limits of risk of soil erosion classes reported from tolerable to severe vary between some countries (see OECD

Figure 7.2. **Trends in agricultural land area classified as having moderate to severe water erosion risk, OECD countries, 1990-2010**

Review of water erosion greater than 11 tonnes/hectare/year of soil loss as a percentage share of total agricultural land area

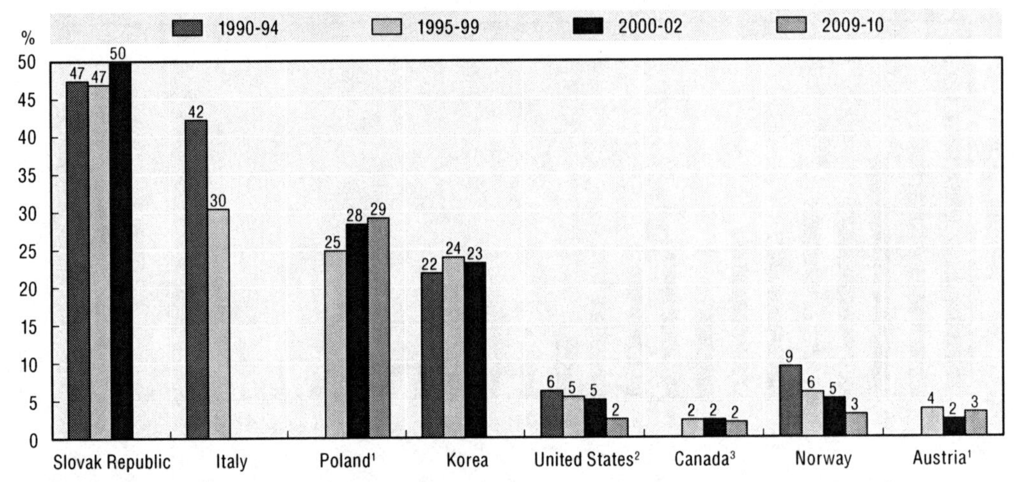

Note: Countries are ranked in terms of highest to lowest % share of agricultural land at risk to water erosion.

1. Average 2000-02 for Austria and Poland refer to 2003-04.
2. Average 2009-10 for United States refer to 2007-08.
3. Average 2009-10 for Canada refer to 2005-06.

Source: OECD (2008), *Environmental Performance of Agriculture in OECD Countries Since 1990*, *www.oecd.org/agriculture*; Joint Research Centre, European Union; Unpublished Estimates Pan-European RUSLE Model JRC, 2011; and national sources.

StatLink http://dx.doi.org/10.1787/888932792901

website database), but, a standardised scale has been used by OECD to present these data. Agricultural soils can "tolerate" a certain amount of erosion without adversely impacting on long-term productivity because new soil is constantly being formed to replace losses.

The tolerable limit varies between different soil depths, types and agro-climatic conditions, but typically ranges from 1 tonne/hectare/year on shallow sandy soils to 6 tonnes/hectare/year on deeper well-developed soils. OECD's scale of soil erosion risk categories ranges from tolerable erosion (less than 6 tonnes/hectare/year) through low, moderate, high and finally to severe erosion (greater than 33 tonnes/hectare/year). However, not all countries use these class limits as some consider tolerable erosion as less than 4 tonnes/hectare/year (e.g. the **Netherlands**, and the **Czech** and **Slovak Republics**). The figures in this chapter, to standardise the presentation between countries, show the area at risk to water and wind erosion above 11 tonnes/hectare/year (i.e. moderate to severe erosion risk).

While the models used to generate soil erosion indicators generate reasonably accurate national trends, they are subject to limitations which need to be taken into account when interpreting the indicators (Eilers et al., 2010; Eurostat, OECD, 2003; and 2008). The underlying datasets used by different national models vary in quality and spatial coverage and, hence, affect the results derived by the models. Also the models do not take into account all the different soil erosion phenomena (e.g. stream channel or gully erosion) nor some erosion control practices, such as grassed waterways and winter cover crops.

Changes in agricultural land cover and use (Chapters 3 and 13), farm production intensity, and management practices and systems are the key driving forces covered by the soil erosion indicators which describe the state (or risk) of on-farm erosion. These indicators are useful tools for policy makers as they provide an assessment of the long-term environmental sustainability of management practices and the effectiveness of soil

Figure 7.3. **Agricultural land area classified as having moderate to severe wind erosion risk, OECD countries, 1995-2010**

Risk of wind erosion greater than 11 tonnes/hectares/year of soil loss, as a percentage share of agricultural land area

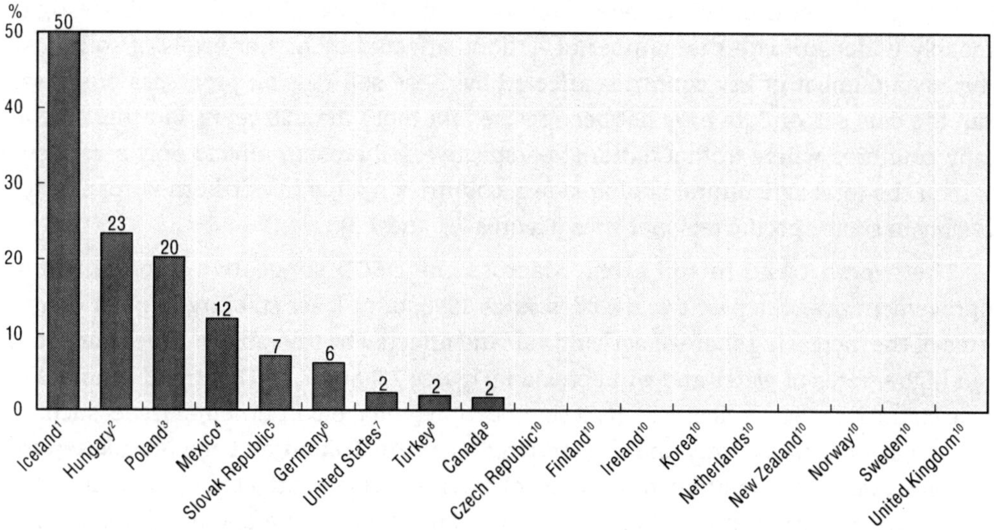

Note: Countries are ranked from highest to lowest % share of agricultural land at risk to wind erosion.

1. Late 1990s, permanent grassland only, comprising 95% of the total severe erosion not classified by soil loss but by farmland.
2. Data for 1995-99.
3. Data shows agricultural land covered by all wind erosion risk categories from tolerable to high erosion risk, for 2009-10.
4. Data for 2000-02.
5. Data for 2003-04.
6. Data refer to year 1996.
7. Data for 2007.
8. Share of agricultural land of risk to elevated erosion rates, but t/ha/y not specified.
9. Data for 2005-06, for cultivated agricultural land.
10. These countries report that the risk of moderate to severe wind erosion was very limited between zero and less than 0.5% of the total agricultural land area.

Source: OECD (2008), *Environmental Performance of Agriculture in OECD Countries Since 1990, www.oecd.org/agriculture;* OECD Agri-Environmental Indicators Questionnaire, unpublished; and national sources.

StatLink http://dx.doi.org/10.1787/888932792920

conservation measures. They can also be related to a range of soil quality issues including the loss of soil organic matter and soil biodiversity.

Measurability and data quality

Some OECD countries have well established soil monitoring systems (e.g. the **United States**) that provide field observations to directly validate national risk estimates. Other OECD countries are at an earlier stage of implementing similar field measurement systems (e.g. **Australia**); while others, including several **European Union** countries, are in the process of designing such monitoring systems (Eurostat and Joint Research Centre).

A number of countries generate national soil erosion data on a regular basis (every 5 years, for example **Canada**, the **United States**), while for the **European Union** the Joint Research Centre has recently generated EU wide soil erosion data based on a harmonised specific model (Joint Research Centre). But for a number of countries where soil erosion degradation (water and/or wind erosion) is a widespread concern, there is little or no regular updating of national soil erosion monitoring (e.g. **Australia**, **New Zealand**, **Portugal**, **Spain** and **Turkey**).

7.3. Main trends

Over 20% of the agricultural land area is affected by moderate to severe soil erosion from water in almost a third of OECD member countries, although far fewer countries suffer a similar level of soil erosion from the wind (Figures 7.1 and 7.3). These figures probably underestimate the number of nations affected by higher levels of soil erosion rates, as a number of key countries affected by these soil erosion processes are missing from the data set, or data have not been revised for more than 20 years. But there are also many countries where from a national perspective soil erosion affects only a very small share of the total agricultural land area (e.g. countries mainly in northern Europe), except possibly in some specific regional sites (Figures 7.1 and 7.3).

The overall trend in soil erosion across the OECD suggests one of continuing improvement over the past two decades since 1990, or at least stability in most cases, in terms of the increasing share of agricultural land affected by tolerable or lower rates rather than higher rates of water and wind erosion (Figures 7.2 and 7.4). This trend is notable for those countries where soil erosion is a significant regional environmental issue, such as in some areas of **Canada**, **Italy** and the **United States**. For **Poland** and the **Slovak Republic**, there has been a deterioration in the share of agricultural land affected by erosion, but this reflects the reduction in the total agricultural land area as the actual area affected by erosion declined in both countries (Figures 7.2 and 7.4).

Figure 7.4. **Trends in agricultural land area classified as having moderate to severe wind erosion risk, OECD countries, 1991-2010**

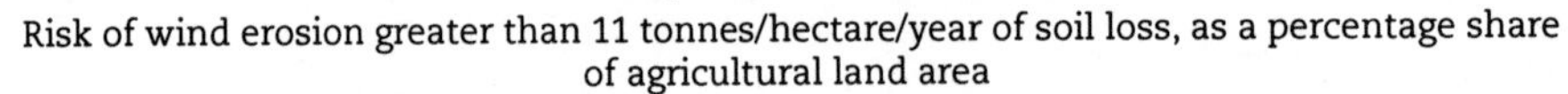
Risk of wind erosion greater than 11 tonnes/hectare/year of soil loss, as a percentage share of agricultural land area

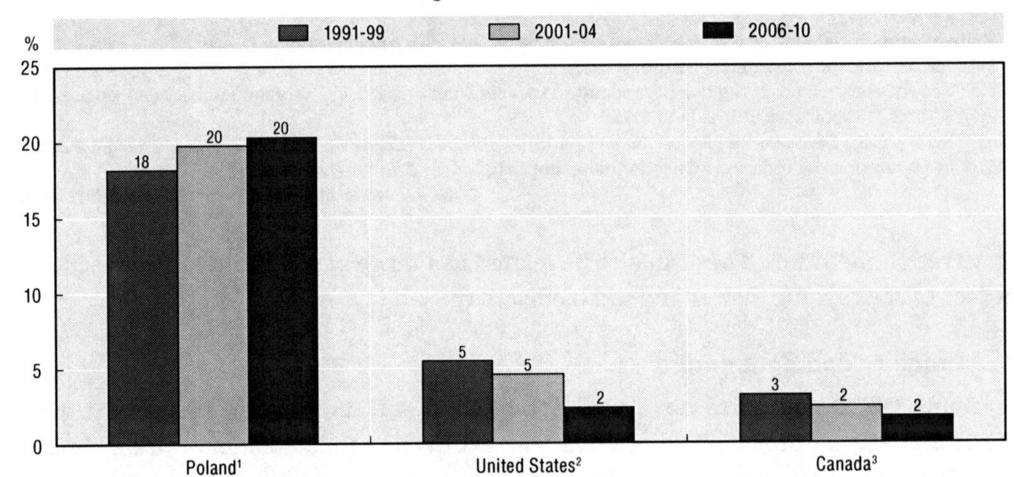

Note: Countries are ranked from highest to lowest % share of agricultural land at risk to wind erosion.

1. Data for 1995-99, 2003-04 and 2009-10.
2. Data for 1992, 2001 and 2007.
3. Data for cultivated land, for 1991, 2001 and 2006.

Source: OECD (2008), *Environmental Performance of Agriculture in OECD Countries Since 1990*, *www.oecd.org/agriculture*; OECD Agri-Environmental Indicators Questionnaire, unpublished; and national sources.

StatLink http://dx.doi.org/10.1787/888932792939

Improvement in reducing the agricultural land area susceptible to a high risk of erosion is mainly linked to both the increased uptake of soil conservation practices, such as the adoption of reduced or no tillage improving soil organic matter content and reducing the number of days the soil is exposed with no vegetative cover, but also the conversion of agricultural land to pasture and forestry in some cases. Trends in soil conservation management and land use/cover changes can provide indirect evidence of likely changes in the areas susceptible to soil erosion. In **Australia**, for example, where national data on soil erosion are limited, the rapid increase in agricultural land under conservation tillage practices, from 20% in 1996 to 75% by 2010, suggest improvements in reducing soil erosion and improving productivity (State of the Environment 2011 Committee). For other countries where data on soil erosion are limited or non-existent, but where soil erosion is an important issue, an increase in the area of pasture (Figure 13.5), might indicate that risks of soil erosion could be diminishing, for example, in **Portugal** and **Turkey**.

Where risks of erosion still remain a concern this is largely attributed to the: continued cultivation of fragile and marginal soils; overgrazing of pasture, especially in hilly/mountainous areas; and the poor uptake of soil conservation practices. Also, in some regions soil erosion is being aggravated by the increasing incidence and severity of droughts and/or heavy rainfall events, most likely linked to climate change (e.g. **Australia**, **Italy** and **Spain**) and in some countries clearing of native vegetation and forests (e.g. **Mexico** and **Turkey**). Land clearing has also been an issue in **Australia**, with around 1 million hectares cleared annually over the decade to 2010, although by 2010 the extent of land clearing was balanced by the extent of regrowth (State of the Environment 2011 Committee).

Soil erosion can originate from a number of economic activities (e.g. forestry, construction, off-road vehicle use) and natural events (e.g. fire, flooding and droughts). In most cases, however, the major share of soil erosion is accounted for by agricultural activities. In general, cultivated arable and permanent crops (e.g. orchards) are more susceptible to higher levels of soil erosion compared to pasture areas. This is because land under pasture is usually covered with vegetative growth all year. In the **United States**, for example, of agricultural land in the moderate to severe water erosion risk classes, arable and permanent cropland accounted for 91% of the total in 2007 (OECD website database). However, where pasture is located on fragile soils with steep topography and subject to intensive grazing, problems of soil erosion can be more acute than on cultivated land, for example, in **Italy**, **New Zealand** and the **United Kingdom** (OECD, 2008).

References

Eilers, W., R. MacKay, L. Graham and A. Lefebvre (eds.) (2010), "Environmental Sustainability of Canadian Agriculture", *Agri-Environmental Indicator Report Series*, Report No. 3, Agriculture and Agri-Food Canada, Ottawa, Canada, *http://publications.gc.ca/collections/collection_2011/agr/A22-201-2010-eng.pdf*.

Eurostat (European Commission Statistical Office), website on agri-environmental indicators, *http://epp.eurostat.ec.europa.eu/portal/page/portal/agri_environmental_indicators/introduction*.

Joint Research Centre of the European Commission website concerning soils, see *http://eusoils.jrc.ec.europa.eu/Library/Themes/Contamination/Irena.html*.

OECD (2008), *Environmental Performance of Agriculture in OECD Countries Since 1990*, OECD Publishing, *www.oecd.org/tad/sustainable-agriculture/agri-environmentalindicators.htm*.

OECD (2003), *Agricultural Impacts on Soil Erosion and Soil Biodiversity: Developing Indicators for Policy Analysis*, OECD Publishing, *www.oecd.org/agr/env/indicators.htm*.

State of the Environment 2011 Committee (2011), *Australia State of the Environment 2011*, Department of Sustainability, Environment, Water, Population and Communities, Australian Government, Canberra, *www.environment.gov.au/soe/2011/index.html*.

Chapter 8

Water resource withdrawals, irrigated area, and irrigation water application rates

This chapter reviews the environmental performances of agriculture in OECD countries related to water resource withdrawals, irrigated area, and irrigation water application rates. It provides a description of the policy context (issues and main challenges), definitions for the agri-environmental indicators presented, and elements related to concepts, interpretations, links to other indicators, as well as measurability and data quality. The chapter then describes the main trends of the agri-environmental indicators, using available data covering the period 1990-2010 and based on a set of tables and figures.

8.1. Policy context

The issue

The scope of managing water resources concerns its efficient and equitable allocation to achieve socially, environmentally and economically beneficial outcomes. For agriculture it includes: irrigation to smooth water supply across the production seasons; water management in rainfed agriculture; management of floods, droughts, and drainage; conservation of ecosystems; and meeting cultural and recreational needs linked to water (OECD, 2010a).

Agricultural water resource management covers a wide range of agricultural systems and climatic conditions across OECD countries. For most countries agriculture is largely rainfed, but in areas susceptible to variable precipitation, irrigation is used to supplement rainfall, mainly drawing on freshwater from surface and groundwater sources (i.e. shallow wells and deep aquifers), and to a lesser extent recycled wastewater and desalinated water. Water resource management in agriculture also operates in a highly diverse set of political, cultural, legal and institutional contexts, encompassing a range of areas of public policy: agriculture, water, environment, energy, fiscal, economic, social and regional.

Main challenges

All OECD countries have policy strategies to address broad water management goals covering water resources, quality and ecosystems. But in terms of the more specific objectives for managing water resources in agriculture, countries largely share a common strategic vision to (OECD, 2010a):

1. establish a long term plan for the sustainable management of water resources in agriculture taking into account climate change impacts, including protection from flood and drought risks;
2. contribute to raising agricultural incomes and achieving broader rural development goals;
3. protect ecosystems on agricultural land or affected by farming activities;
4. balance agricultural water withdrawals with environmental needs, especially maintaining minimum flow levels in rivers and lakes and ensuring sustainable use of groundwater resources (i.e. both shallow wells and deep aquifers); and
5. improve water resource withdrawal efficiency, management and technologies on-farm and ensure the financing to maintain and upgrade the infrastructure supplying water to farms (and other water consumers).

8.2. Indicators

Definitions

The indicators related to agricultural freshwater resources include changes in:

- Agricultural freshwater withdrawals.
- Irrigated land area.
- Irrigation water application rate – megalitres of water applied per hectare of irrigated land.

Concepts, interpretation, limitations and links to other indicators

In many OECD countries there is growing competition for water resources between industry, household consumers, agriculture and the environment (i.e. aquatic ecosystems). The demand for water is also affecting aquatic ecosystems, particularly where water extraction is in excess of minimum environmental needs for rivers, lakes and wetland habitats. However, some OECD countries possess abundant water resources and, as a result, do not consider water availability to be a significant environmental issue in terms of resource protection. There are also important social issues concerning water, such as access for recreational needs (e.g. swimming, boating, fishing) and the aesthetic value of waterscapes. Moreover, in some societies water has a significant cultural and spiritual value, for example, for the indigenous peoples of **Australia**, **Canada** and **New Zealand**, and for much of the population in **Japan** and **Korea** (OECD, 2004).

The indicators in this chapter provide information on trends in agricultural freshwater withdrawals and reveal the importance of the sector in total national freshwater withdrawals. In addition, the indicators include monitoring changes in irrigated area as a share of the total agricultural area, and the implications of this change for water withdrawals efficiency in agriculture, measured in terms of the megalitres of water applied per hectare irrigated. As the irrigation sub-sector accounts for the major share of agricultural water withdrawals in most countries, it is a key driving force that affects agricultural water withdrawals.

The indicators related to agriculture water resources have a number of limitations which need to be taken into account when examining absolute levels, trends and comparing countries, as discussed below (discussion of the limitations of the water application indicator are discussed in the following section):

1. Methods of collecting and calculating the data vary across and within countries and are also subject to errors of measurement. Sources of data for irrigation freshwater withdrawals include sample surveys of irrigators, and are sometimes estimated using information on irrigated crop acreages along with specific crop water-consumption coefficients or irrigation-system application rates. In other cases irrigation water withdrawal data may reflect water allocations, which may differ substantially from actual withdrawals depending on annual climatic conditions. These estimates may or may not include adjustments for climatic variables, system efficiencies, conveyance losses, and other irrigation practices such as pre-irrigation (Kenny et al., 2009). The reliability of these estimates where surveys are used are also subject to sampling errors, because not all farms are included in the surveys. However, evidence from **Australia** (Australian Bureau of Statistics, 2012) reveals that estimates of sampling errors for these indicators are very low (i.e. below 5%).
2. The coverage of data can vary between countries. It is assumed in this chapter that total and agricultural water withdrawals covers only freshwater sources, but in some regions, recycled wastewater and desalinated water is used by agriculture. In cases where agriculture uses these water sources, they are usually in very small volumes and not included in agricultural water withdrawal calculations. Saline water use in the **United States**, for example, was 15% of total water withdrawals in 2005, but nearly all of this was used by the power sector (Kenny et al., 2009). **Israel** is a notable exception, with 54% of water resources allocated to agriculture derived from recycled effluent and desalinated water in 2008, compared to a share of 33% in 1998 (OECD, 2010b).

3. There are practical difficulties in accurately measuring agricultural water withdrawals. In particular, the extent of groundwater withdrawals on farm, either from shallow wells or deep aquifers, can be difficult to monitor, especially as in most cases groundwater withdrawals are not metered. An additional complication is that under some systems, agriculture has the potential to recharge groundwater. Estimates of return flows of water from irrigation systems, for withdrawal by other users further downstream in a catchment, are also subject to considerable uncertainty.

As environmental driving forces, the agricultural water withdrawals and irrigated area indicators are linked to the state of (changes in) groundwater reserves and competition over water resources with other major water consumers. Responses to these changes in the sustainability of water withdrawals are revealed through the uptake of more efficient irrigation management technologies and practices and resulting improvements in water application rates per hectare of land irrigated.

Measurability and data quality

The term "agricultural water withdrawals" used in this chapter refers to "water abstractions" for irrigation and other agricultural withdrawals (such as for livestock) from rivers, lakes, reservoirs and groundwater (shallow wells and deep aquifers), and "return flows" from irrigation, but excludes precipitation directly onto agricultural land. "Water withdrawal" is different from "water consumption", which relates to water depleted and not available for reuse.

Canadian agriculture's withdrawals of water resources provides an illustrative numerical example of the application of these terms. Agriculture in Canada withdraws from total available renewable freshwater sources (mainly from precipitation and groundwater sources but also melting glaciers, etc.) 6% of Canada's overall freshwater resources (Figure 8.1). Agriculture consumes (does not return to the water system) 70-80% of the water it withdraws to make it the leading consumer of water in Canada (about 70% of total consumption) (OECD, 2010a).

Irrigation freshwater withdrawals, for most countries, usually include water that is applied by an irrigation system to sustain plant growth, including arable and horticultural crops as well as pasture. Irrigation also includes water that is applied for pre-irrigation, frost protection, application of chemicals, weed control, field preparation, crop cooling, harvesting, dust suppression, leaching salts from the root zone, and water lost in conveyance (Kenny et al., 2009). For some countries, irrigation may cover golf courses, parks, and other non-agricultural uses, and include self-supplied withdrawals and deliveries from private or government companies, districts, and cooperatives.

Most OECD countries have incomplete series of data for total and agricultural water withdrawals and irrigated area (see the OECD website database). This is because, in part, these data are usually not calculated annually, but derived from five- or even ten-year surveys. Other limitations that compromise the quality of water resource data are discussed above.

8.3. Main trends

Overall the key trends in total OECD agriculture's withdrawal of freshwater resources over two decades from 1990 to 2010, include (Figures 8.1, 8.2 and 8.3):

- withdrawals of freshwater resources by agriculture have declined over the decade of the 2000s by -0.3 per annum, compared to an increase of +0.3 per annum over the 1990s;

Figure 8.1. **Agricultural water withdrawals, OECD countries, 1990-2010**

Share of agriculture in total water withdrawals (%) 2008-10

0 25 50 75 100
%

	Total agricultural freshwater withdrawals			Change in total agricultural freshwater withdrawals		Change in total freshwater withdrawals		Share of agriculture in total freshwater withdrawals
	1990-92[1]	1998-2000[2]	2008-10[3]	1990-92 to 1998-2000	1998-2000 to 2008-10	1990-92 to 1998-2000	1998-2000 to 2008-10	2008-10
	million m^3			Average annual % change		Average annual % change		%
Greece	7 733	8 809	8 457	1.3	-0.7	2.1	-0.3	89
Turkey	26 240	35 192	36 274	3.7	0.3	3.2	0.7	85
Mexico	62 500	61 385	61 190	-0.4	0.0	0.8	0.5	77
Chile[4]	16 273	19 268	16 619	1.9	-2.1	2.1	-1.0	74
Japan	58 642	57 930	54 650	-0.2	-0.7	-0.1	-0.7	66
Portugal	6 569	6 308	4 392	-0.4	-5.0	0.3	-3.1	63
New Zealand	..	2 818	3 207	..	3.3	..	1.5	62
Spain	23 700	25 008	20 255	0.7	-2.6	0.1	-1.5	61
Israel	997	1 256	1 108	2.9	-1.4	3.3	-0.5	56
Australia	14 498	15 033	7 359	0.6	-7.6	1.3	-5.2	52
Korea	14 700	15 810	16 000	0.9	0.1	3.6	0.2	47
OECD[5]	**441 503**	**449 848**	**427 663**	**0.2**	**-0.5**	**0.3**	**-0.3**	**44**
Iceland	70	70	70	0.0	0.0	-0.4	0.4	42
United States	195 200	191 555	192 485	-0.2	0.1	0.2	0.2	40
Denmark	383	194	201	-8.2	0.4	-4.9	-1.4	31
Norway	..	769	807	..	0.7	..	3.3	27
EU15[6]	**45 615**	**46 612**	**38 134**	**0.3**	**-2.0**	**-0.8**	**-1.0**	**26**
United Kingdom	1 347	2 015	1 294	5.2	-5.4	-1.0	-2.5	15
Poland	1 527	1 035	1 154	-4.7	1.1	-1.9	-0.6	10
France	4 901	3 871	3 218	-2.9	-2.0	-2.2	0.2	10
Canada	3 991	4 104	2 257	0.6	-5.3	-1.3	-1.1	6
Hungary	968	523	305	-7.4	-6.5	-1.5	-1.7	6
Sweden	169	150	126	-1.5	-3.5	-1.2	-0.3	5
Slovak Republic	188	63	21	-12.8	-11.3	-5.3	-6.3	3
Czech Republic	93	13	37	-22.1	11.2	-6.0	-0.1	2
Netherlands	184	85	75	-7.4	-2.1	1.1	3.2	1
Belgium	13	28	36	22.3	3.6	-0.6	-2.2	1
Estonia[4]	..	..	4	..	..	-7.8	-1.0	0.3
Germany[4]	616	142	80	-18.9	-10.9	-1.6	-2.8	0.2
Luxembourg	0	0	0	-5.4	-6.1	1.6	-2.1	0.2
Slovenia	4	5	2	3.3	-18.4	..	1.2	0.2
Austria	100	100	..	0.0	..	-0.4	-0.5	..
Finland	20	50	50	12.1	0.0	-0.3	..	..
Ireland	179	..	..	..	..	..	..	..
Italy[4]	..	20 865	16 800	..	-2.0	..	..	..

Notes: Agricultural water withdrawals is defined as freshwater resources for irrigation and other agricultural uses (e.g. livestock operations), including water abstractions from rivers, lakes, reservoirs, groundwater and "return flows" from irrigation, but excludes precipitation directly onto agricultural land. In compiling total freshwater withdrawals, water returns abstracted by another water use downstream are counted again.
Countries are ranked in descending order according to share of agriculture in total freshwater withdrawals.
The statistical data for Israel are supplied by and under the responsibility of the relevant Israeli authorities. The use of such data by the OECD is without prejudice to the status of the Golan Heights, East Jerusalem and Israeli settlements in the West Bank under the terms of international law.

1. Data for 1990-92 average equal to the 1994-95 average for Mexico; the 1994-96 average for Australia and Belgium; the year 1991 for Canada, Netherlands and Spain; the year 1990 for Chile; two-years 1990 and 1992 average for the Czech Republic and Hungary; the 1990-91 average for Denmark; the 1992-94 average for Iceland; the year 1995 for Germany and Luxembourg; the year 1994 for Ireland; and the year 1990 for Japan, Korea, Portugal and United States.
2. Data for 1998-2000 average equal to the year 2001 for Australia and the Netherlands; the 1998-99 average for Austria; the year 1999 for Chile, Finland, Luxembourg and Norway; the year 2006 for New Zealand; the year 1996 for Canada; the year 2002 for Germany; the year 1998 for Italy and Korea; the 2000-02 average for Greece; the year 2000 for Portugal and United States; and the 2002-04 average for Slovenia.
3. Data for 2008-10 average equal to the year 2010 for Australia, Luxembourg and New Zealand; the 2005-07 average for Belgium and Norway; the year 2007 for Canada, Germany, Korea and Portugal; the year 2006 for Chile; the 2007-09 average for Denmark, France, Israel, Mexico, the Slovak Republic and Turkey; the year 2009 for Estonia and Italy; the 2006-07 average for Greece; the 2006-08 average for Hungary, Japan, Spain and United Kingdom; the 2007-08 average for Netherlands; the 2003-05 average for Finland, Iceland and Sweden; and the year 2005 for United States.
4. For Chile, Estonia, Germany and Italy, data for irrigation water are used because data for agricultural water withdrawals are not available.
5. Due to unavailable data, OECD does not include Austria, Estonia, Finland, Ireland, Italy, New Zealand, Norway, Slovenia and Switzerland.
6. Due to unavailable data, EU15 does not include Austria, Finland, Ireland, and Italy.

Source: OECD (2008), *Environmental Performance of Agriculture in OECD Countries Since 1990, www.oecd.org/tad/sustainable-agriculture/agri-environmentalindicators.htm*; OECD (2008), *OECD Environmental Data Compendium, www.oecd.org/environment*; and national sources.

StatLink http://dx.doi.org/10.1787/888932792958

Figure 8.2. **Irrigated area, OECD countries, 1990-2010**

Share of irrigated area in total agricultural area (%) 2008-10

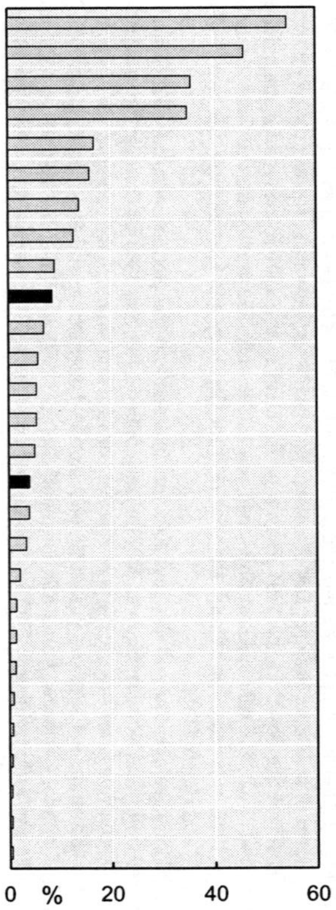

	Irrigated area			Change in irrigated area		Change in total agricultural area		Share of irrigated area in total agricultural area	Share of irrigation freshwater withdrawal in total agricultural freshwater withdrawal
	1990-92[1]	1998-2000[2]	2008-10[3]	1990-92 to 1998-2000[1,2]	1998-2000 to 2008-10[2,3]	1990-92 to 1998-2000[1,2]	1998-2000 to 2008-10[2,3]	2008-10[4]	2008-10[5]
	1000 hectares			% per annum		% per annum		%	%
Japan	2 824	2 660	2 506	-0.7	-0.6	-0.8	-0.5	54	100
Korea	984	880	829	-1.4	-0.7	-1.4	-0.9	46	100
Israel	191	191	185	0.0	-0.4	-0.3	-0.9	36	--
Greece	1 229	1 432	1 423	1.9	-0.1	-0.2	1.9	36	100
Denmark	433	451	449	0.5	-0.1	0.4	-0.8	17	100
Italy	2 707	2 699	2 199	0.0	-2.0	-1.5	-1.5	16	--
Spain	3 398	3 694	3 351	1.0	-1.4	-1.5	-1.0	13	100
Portugal	631	606	469	-0.4	-2.5	-0.4	-0.5	13	100
Turkey	3 329	3 093	3 506	-0.9	1.3	-0.5	0.0	9	100
EU15[6]	**10 566**	**11 125**	**9 880**	**0.6**	**-1.2**	**-0.7**	**-0.5**	**9**	**96**
Chile	--	1 067	1 094	--	0.2	-0.5	0.4	7	100
New Zealand	--	471	650	--	4.1	-0.4	-1.2	6	87
United States	21 067	22 622	22 567	0.9	0.0	-0.3	0.0	5	92
Mexico	5 353	4 944	5 563	-1.0	1.3	0.2	-0.4	5	97
France	1 485	1 575	1 489	0.6	-0.8	-0.2	-0.2	5	100
OECD[7]	**47 829**	**49 144**	**47 579**	**0.3**	**-0.3**	**-0.3**	**-0.5**	**4**	**95**
Norway	--	44	40	--	-1.3	0.5	-0.3	4	14
Switzerland	--	43	36	--	-2.8	0.1	-0.2	3	--
Germany	482	485	373	0.1	-2.6	-0.1	-0.2	2	100
Austria	46	37	44	-2.9	4.8	-0.3	-0.7	1	82
Hungary	205	84	77	-10.6	-0.8	-0.6	-0.6	1	21
Slovak Republic	250	178	25	-6.6	-17.8	0.1	-2.3	1	54
Canada	719	765	529	0.8	-3.3	0.1	0.2	1	74
Slovenia	--	2	3	--	5.6	-1.5	-0.4	1	100
Czech Republic	24	24	19	0.0	-2.0	0.0	-1.9	0.5	54
Poland	259	103	76	-10.9	-3.0	-0.3	-1.3	0.5	8
United Kingdom	156	147	83	-0.9	-6.2	-0.4	-0.2	0.5	6
Australia	2 057	2 476	1 817	3.8	-4.3	-0.2	-1.1	0.4	90
Belgium	18	40	--	10.5	--	0.2	-0.2	--	100
Netherlands	557	565	--	0.2	--	-0.2	-0.3	--	51
Sweden	115	115	--	0.0	--	-1.3	0.1	--	68

Figure 8.2. **Irrigated area, OECD countries, 1990-2010** *(cont.)*

Notes: Countries are ranked in descending order according to share of irrigated area in total agricultural area.

The statistical data for Israel are supplied by and under the responsibility of the relevant Israeli authorities. The use of such data by the OECD is without prejudice to the status of the Golan Heights, East Jerusalem and Israeli settlements in the West Bank under the terms of international law.

1. For irrigated area, data for 1990-92 average equal to the year 1997 for Australia; the year 1993 for the Czech Republic; the year 1990 for France and Portugal; the 1993-95 average for the Slovak Republic; and the year 1995 for Austria and United Kingdom. For total agricultural area, data for 1990-92 average equal to the 1991-93 average for Slovenia and the year 1990 for Greece and Switzerland.
2. For irrigated area, data for 1998-2000 average equal to the 2001-03 average for Australia; the 2003-04 average for Austria; the year 1997 for Chile; the year 2000 for France and United Kingdom; the 1998-99 average for Greece; the 2002-03 average for New Zealand; the year 2004 for Norway; the year 1999 for Portugal; the 2002-04 average for Slovenia; and the 2004-05 average for Switzerland. For total agricultural area, data for 1998-2000 average equal to the 1999-2001 average for Austria and the year 2000 for Greece.
3. For irrigated area, data for 2008-10 average equal to the year 2007 for Austria, Chile and France; the year 2009 for Germany, Italy and Portugal; the year 2010 for Canada, the Czech Republic, New Zealand, Norway, Switzerland and United Kingdom; the year 2004 for Denmark; the 2007-09 average for Greece, Korea and Mexico; the 2006-08 average for Israel; the 2005-07 average for Spain; and the 2007-08 average for United States. For total agricultural area, data for 2008-10 average equal to the 2007-08 average for Italy; the 2007-09 average for Austria, Canada, Chile, Denmark, Israel, Korea and Mexico; and the year 2007 for Greece.
4. For the share of irrigated area in the total agricultural area, the 2008-10 ratio refers to the year 2007 for Austria, Chile, France and Greece; the year 2010 for irrigated area and the year 2009 for total agricultural area for Canada; the year 2010 for Czech Republic, New Zealand, Norway and Switzerland; the year 2004 for Denmark; the year 2009 for Germany; the 2006-08 average for Israel; the year 2009 for irrigated area and the year 2008 for total agricultural area for Italy; the 2007-09 average for Korea and Mexico; the 2005-07 average for Spain; and the 2007-08 average for United States.
5. For irrigation freshwater withdrawal and total agricultural freshwater withdrawal, data for 2008-10 average equal to the 2002-03 average for Austria; the 2005-07 average for Belgium and Norway; two-year 2005 and 2007 average for Canada; the year 2006 for Chile; the 2007-09 average for Denmark, France, Greece, Netherlands and Slovak Republic; the year 2007 for Germany; the 2004-06 average for Hungary; the 2003-05 average for Sweden; the 2006-08 average for Japan, Mexico, Spain and United Kingdom; the year 2002 for Korea; the year 2010 for New Zealand; two-year 2007 and 2009 average for Portugal; and the year 2005 for United States.
6. Due to unavailable data, EU15 does not include for irrigated area and total agricultural area: Belgium, Finland, Ireland, Luxembourg, Netherlands and Sweden; for irrigation freshwater withdrawal and total agricultural freshwater withdrawal: Finland, Ireland, Italy and Luxembourg.
7. Due to unavailable data, OECD does not include for irrigated area and total agricultural area: Belgium, Chile, Estonia, Finland, Iceland, Ireland, Luxembourg, Netherlands, New Zealand, Norway, Slovenia, Sweden and Switzerland; for irrigation freshwater withdrawal and total agricultural freshwater withdrawal: Estonia, Finland, Iceland, Ireland, Israel, Italy, Luxembourg and Switzerland.

Source: OECD (2008), *Environmental Performance of Agriculture in OECD Countries Since 1990*, *www.oecd.org/tad/sustainable-agriculture/agri-environmentalindicators.htm*; OECD (2008), *OECD Environmental Data Compendium*, *www.oecd.org/environment*; Statistical Office of the European Union (EUROSTAT), *http://epp.eurostat.ec.europa.eu*; and national sources.

StatLink http://dx.doi.org/10.1787/888932792977

Figure 8.3. **Irrigation water application rates, OECD countries, 1990-2010**

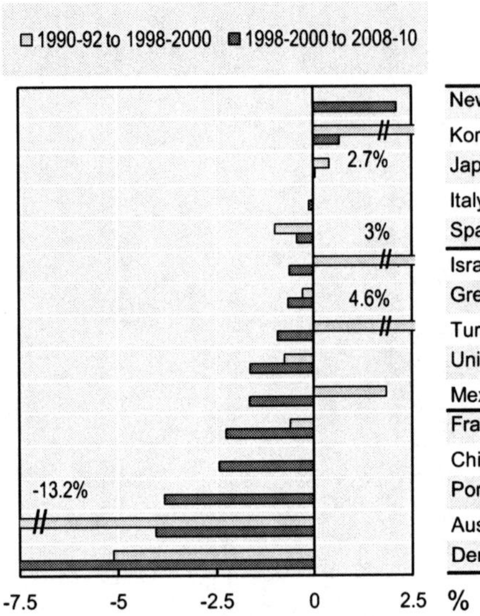

	Irrigation water application rates[1] (Megalitres per hectare of irrigated land)			Average annual % change (% per annum)	
	1990-92[2]	1998-2000[3]	2008-10[4]	1990-92 to 1998-2000	1998-2000 to 2008-10
New Zealand	..	3.4	4.3	..	2.1
Korea	14.3	17.6	18.2	2.7	0.7
Japan	20.6	21.5	21.6	0.4	0.1
Italy	..	7.7	7.6	..	-0.1
Spain	7.0	6.5	6.3	-1.0	-0.4
Israel	5.2	6.6	6.2	3.0	-0.6
Greece	6.3	6.1	5.8	-0.3	-0.7
Turkey	7.9	11.4	10.3	4.6	-0.9
United States	9.1	8.4	7.7	-0.8	-1.7
Mexico	11.4	12.2	10.7	1.8	-1.7
France	3.3	3.1	2.6	-0.6	-2.3
Chile	..	18.1	15.2	..	-2.4
Portugal	10.4	10.4	7.3	0.0	-3.8
Australia	8.7	4.9	3.6	-13.2	-4.0
Denmark	0.9	0.7	0.4	-5.1	-7.5

Notes: The figures only include those OECD countries where irrigation area exceeds 5% of total agricultural area, with the exception of Australia where irrigated agriculture is important (irrigation accounts for over 50% of total freshwater withdrawals) but is less than 5% of agricultural land because of the large area under pasture.
Countries are ranked in descending order according to average annual % change 1998-2000 to 2008-10.
Data for Israel refer to agricultural freshwater withdrawals. The statistical data for Israel are supplied by and under the responsibility of the relevant Israeli authorities. The use of such data by the OECD is without prejudice to the status of the Golan Heights, East Jerusalem and Israeli settlements in the West Bank under the terms of international law.

1. Irrigation water application rates are calculated as the quantities of irrigated freshwater withdrawals divided by the irrigated area (Figure 8.2).
2. Data for 1990-92 average equal to the year 1997 for Australia; the 1990-91 average for Denmark; the year 1990 for France, Japan, Korea, Portugal and United States; the 1990-92 average for Greece; the 1994-95 average for Mexico; and the year 1991 for Spain.
3. Data for 1998-2000 average equal to the year 2001 for Australia; the year 1999 for irrigation water withdrawals and the year 1997 for irrigation area for Chile; the 1995-96 average for Denmark; the year 2000 for France; the 2000-02 average for Greece; the year 1998 for Italy; the 2000-01 average for Japan; the 1997-98 average for Korea; the 1998-2000 average for Mexico, Spain and Turkey; the year 1999 for irrigation water withdrawals and the year 2002 for irrigation area for New Zealand; the year 2000 for irrigation water withdrawals and the year 1999 for irrigation area for Portugal; and the year 2000 for United States.
4. Data for 2008-10 average equal to the 2007-09 average for Greece; the year 2006 for irrigation water withdrawals and the year 2007 for irrigation area for Chile; the 2002-04 average for Denmark; the year 2007 for France; the 2006-08 average for Israel and Mexico; the 2007-08 average for Japan; the 2002-03 average for Korea; the year 2010 for New Zealand; the year 2009 for Italy and Portugal; the 2005-07 average for Spain; and the year 2005 for irrigation water withdrawals and the year 2007 for irrigation area for United States.

Source: OECD (2008), *Environmental Performance of Agriculture in OECD Countries Since 1990*, *www.oecd.org/tad/sustainable-agriculture/agri-environmentalindicators.htm*; OECD (2008), *OECD Environmental Data Compendium*, *www.oecd.org/environment*; and national sources.

StatLink http://dx.doi.org/10.1787/888932792996

- agriculture accounts for 44% of total freshwater withdrawals (2008-10), but these shares vary considerably across countries;
- changes in the area irrigated has reflected the trends in agricultural water withdrawals, with a slight increase over the 1990s, but decreasing by -0.4% per annum over the last decade; and
- efficiency of water application on irrigated land improved for most countries over the 2000s (i.e. less water applied per hectare irrigated) compared to a more variable performance over the 1990s.

The overall annual reduction in OECD agricultural water withdrawals during 1998-2000 to 2007-09, suggests that agriculture exerted less pressure on water resources than other water users. This was in contrast to the 1990s when agriculture water withdrawals were increasing. But agriculture remains a major user of water accounting for over 40% of total water withdrawals for nearly a half of OECD member countries (Figure 8.1). Some of the water withdrawn by irrigated agriculture is reused by other downstream users or to meet environmental needs, although there are also losses due to evapotranspiration, pollutant run-off from irrigated farming, and losses to groundwater sources which are no longer economic to pump.

The declining OECD trend in agricultural water withdrawals over the past decade – notably in **Australia**, **Chile**, **EU15**, **Israel** and **Japan** – was mainly driven by a mix of factors, varying between countries, including: a near stable or reduction in the area irrigated; improvements in irrigation water management and technologies efficiency; drought; release of water to meet environmental needs; and slowdown in the growth of agricultural production (except for **Chile** and **Israel**). At the same time, the 2000s saw an increase in the area under irrigation, notably in **New Zealand** (in particular enlarging the area irrigated for dairy pasture land), and **Mexico** and **Turkey** with an expanding irrigated agricultural sub-sector (Figure 8.2).

Agriculture abstracts an increasing share of its water supplies from groundwater. The sector's share in total groundwater utilisation is important for a number of OECD countries where irrigated agriculture is significant, notably, **Denmark**, **Greece**, **Israel**, **Japan**, **Korea**, **Mexico**, **Portugal**, **Spain**, **Turkey** and the **United States** (OECD, 2008; 2010a). Although data are limited, farming is drawing an increasing share of its supplies from groundwater, and agriculture's share in total groundwater withdrawals is also high in many OECD member countries (OECD, 2008).

Over-exploitation of water resources by agriculture in certain areas is damaging ecosystems by reducing water flows below minimum flow (stock) levels in rivers, lakes and wetlands, which is also detrimental to recreational, fishing and cultural uses of these ecosystems. Groundwater withdrawals for irrigation above recharge rates in some regions of notably, **Australia**, **Greece**, **Italy**, **Mexico** and the **United States** is also undermining the economic viability of farming in affected areas. Agriculture is also a major and growing source of groundwater pollution across many countries. This is of particular concern where groundwater provides a major share of drinking water supplies for both human and the farming sector (Chapter 9 and OECD, 2010a).

In some OECD countries ***water stress*** is an issue, which in future could have implications for fresh water withdrawals by agriculture. Water stress is based on the ratio of total water withdrawals (across all uses in the economy, including agriculture) to total annual renewable freshwater availability. For most OECD countries this ratio is low and below 10%, but some countries are experiencing medium water stress above 20%, where water supply and demand needs to be managed to resolve conflicts between competing uses (OECD, 2010a). Countries with a medium water stress (**Belgium**, **Italy**, **Korea**, **Spain**), also have agricultural sectors which account for over 40% of total water withdrawals (except **Belgium**). **Israel** stands out as one of the world's most severely water stressed countries, with a ratio of water withdrawal to annual water availability of around 90% (OECD, 2010b).

In those regions were growing water scarcity is an issue, greater use is being made of ***recycled wastewater and desalinated water*** from seawater and saline aquifers. These sources of water still remain marginal in most OECD countries, although they are important for agriculture in some localities, especially near large population centres (recycled sewage wastewater) and coastal areas (desalinisation), such as beginning to emerge in some OECD Mediterranean countries, for example, **Israel** and **Spain** (OECD, 2010a; 2010b).

Irrigated agriculture provides a major share of the value of farm production and exports for some OECD countries, and supports rural employment in a number of regions. As such irrigated agriculture accounts for a significant share of agricultural water withdrawals, and will continue to play an important role in agricultural production growth in some countries.

Overall the total OECD area irrigated decreased over the 2000s at -0.3% per annum, compared to slight increase over the 1990s (Figure 8.2). The reduction in the area irrigated over the past decade largely reflects decreases in **Australia, Japan** and the **EU15**, especially in **Italy**, **Portugal** and **Spain**, mainly linked to, but varying between countries: reductions in agricultural production; improvements in efficiency with the remaining areas irrigated; and prolonged drought in some regions. Few countries have experienced an expansion in the area irrigated over the past decade, except **Mexico**, **New Zealand** and **Turkey**, closely linked to their expanding agricultural sectors (Figure 3.1).

For many countries the irrigable area (i.e. land with irrigation infrastructure but not necessarily irrigated) is usually much greater than the area actually irrigated (which is shown in Figure 8.2) for any given year. For example, the irrigated area as a share of the irrigable area over 2008-10 was 60% in **France**, 80% in **Italy** and 85% in **Portugal**. This underlines the potential to expand irrigated agriculture, which already for a number of countries generates a high and growing share in total agricultural production value (in excess of 50%) and value of exports (more than 60%), for example, in **Italy**, **Mexico**, **Spain** and the **United States** (crop sales only) (OECD, 2008).

Critical to increasing agricultural production from irrigated land, improving the profitability of irrigated agriculture, and in making water savings in areas of water stress, is increasing the physical (technical) and economic (value of output per unit of water withdrawn) productivity of water withdrawals by irrigators. This is being achieved in many OECD countries through better management and uptake of more efficient technologies, such as drip irrigation, and adoption of other water saving farm practices. In addition, many countries are undertaking agricultural and water policy reforms that seek to transmit the value of supplying water to irrigators by lowering support for water supplied to agriculture, which in some cases has led to allocation of water to higher valued commodities which frequently require less water, such as vines and horticultural crops (OECD, 2010a).

For nearly all OECD countries average water application rates per hectare irrigated decreased over the 2000s compared to a more mixed performance over the 1990s (Figure 8.3). **Australia** stands out as the OECD country making the largest improvement in irrigation water application rates consistently over the two decades from 1990 to 2010 (Figure 8.3). Not only has Australia embraced the adoption of improved irrigation technologies and management practices, but also undertaken major water policy reforms affecting agriculture, including changing water property rights, creating water trading markets and increasing water charges to farmers (OECD, 2010a; Young, 2010). **Israel** has also undertaken significant water policy reforms leading to, in particular, an increase in the charges paid by irrigators for water supplies, which has stimulated a reduction in water application rates per hectare irrigated and led to improvements in irrigation technologies and management (Figure 8.2) (OECD, 2010b).

Other OECD countries are also developing water policy reforms in agriculture, usually as part of economy wide policy reforms, such as under the **EU's** *Water Framework Directive* (OECD, 2010a). In most cases, however, these reforms have yet to significantly reduce irrigators' water application rates, and much of the decrease in water application rates shown in Figure 8.3 have largely been driven over the past decade by improvements in irrigation technologies and management practices (OECD, 2010a). The adoption of drip irrigation, low-pressure sprinkler systems, and other water-saving technologies and practices, are becoming more widespread (OECD, 2008; 2010a). In addition, irrigation water management efficiency in agriculture is being improved, for example, through replacing earthen irrigation channels with concrete linings to reduce losses and upgrading flood irrigation systems (e.g. levelling of fields, neutron probes for soil moisture measurement, and scheduling of irrigation to plant needs).

Despite the adoption of these improved irrigation technologies and management practices, some countries (notably **Mexico** and **Turkey**) continue to experience inefficiencies in irrigation systems, partly explained by losses from the irrigation infrastructure, but also inefficiencies stemming from the lack of irrigator management skills and poor advisory services (OECD, 2008; 2010a). But water policy reforms in both **Mexico** and **Turkey** are beginning to address these deficiencies in managing irrigation systems (Cakmak, 2010; Garrido and Calatrava, 2010; OECD, 2010a).

In reviewing the changes in irrigation water application rates shown in Figure 8.3, it is the trends that are important and not the absolute levels of megalitres applied per hectare of irrigated land. Water application rates can differ greatly both within and between countries, mainly because of: differences in the crops irrigated (e.g. from vines to rice); variations in climate and seasonal rainfall patterns (e.g. monsoonal conditions to arid environments); and differences in the irrigation technologies (e.g. spray guns or micro irrigation) and management practices (e.g. gravity flood irrigation, drip irrigation) used to apply water on irrigated fields.

Under the predominantly monsoonal paddy rice systems of **Japan** and **Korea**, for example, which operate differently to irrigated systems in semi-arid or arid areas, water application rates are significantly higher, around 18-22 megalitres per hectare irrigated compared to an average of about 4-10 megalitres per hectare irrigated for most other irrigation systems (Figure 8.3) (OECD, 2010a). Moreover, for climatic reasons use of irrigation water can be for a limited part of the year, for example, in the more temperate climate of **Denmark** and **France**, compared to longer periods of using irrigation water in the more arid and semi-arid areas of, for example, **Australia**, **Greece**, **Spain** and **Turkey**.

References

Australian Bureau of Statistics (2012), *Water use on Australian Farms, 2010-11*, Canberra, Australia, *www.abs.gov.au/AUSSTATS/abs@.nsf/Lookup/4618.0Main+Features12010-11?OpenDocument*.

Cakmak, E. (2010), *Agricultural Water Pricing: Turkey*, OECD consultant report, available at: *www.oecd.org/water*.

Garrido, A. and J. Calatrava (2010), *Agricultural Water Pricing: EU and Mexico*, OECD consultant report, available at: *www.oecd.org/water*.

Kenny, J.F., N.L. Barber, S.S. Hutson, K.S. Linsey, J.K. Lovelace and M.A. Maupin (2009), *Estimated Use of Water in the United States in 2005*, U.S. Geological Survey Circular 1344, 52 pp., *http://pubs.usgs.gov/circ/1344*.

OECD (2010a), *Sustainable Management of Water Resources in Agriculture*, OECD Publishing, *www.oecd.org/agriculture/water*.

OECD (2010b), *OECD Review of Agricultural Policies: Israel 2010*, OECD Publishing.

OECD (2008), *Environmental Performance of Agriculture in OECD Countries Since 1990*, OECD Publishing, *www.oecd.org/tad/sustainable-agriculture/agri-environmentalindicators.htm*.

OECD (2004), *Agricultural Water Quality and Water Use: Developing Indicators for Policy Analysis*, OECD Publishing, *www.oecd.org/tad/sustainable-agriculture/agri-environmentalindicators.htm*.

Young, M.D. (2010), *Environmental Effectiveness and Economic Efficiency of Water Use in Australia: The Experience of and Lessons from the Australian Water Reform Programme*, OECD consultant report, available at: *www.oecd.org/water*.

Chapter 9

Water quality: Nitrates, phosphorus and pesticides

This chapter reviews the environmental performances of agriculture in OECD countries related to water quality. It provides a description of the policy context (issues and main challenges), definitions for the agri-environmental indicators presented, and elements related to concepts, interpretations, links to other indicators, as well as measurability and data quality. The chapter then describes the main trends of the agri-environmental indicators, using available data covering the period 1990-2010 and based on a set of tables and figures.

9.1. Policy context

The issue

Improving water quality is consistently ranked as a top environmental concern in public opinion surveys across most OECD countries. Over decades, policy actions and major investment in OECD countries has helped to drastically reduce water pollution from urban centres, industry and sewage treatment works, with substantial gains for the economy, human health, environment and social values linked to water. In the light of this success focus has now switched in many countries to addressing agricultural water pollution. This is because agricultural water pollution principally originates from farms spread across the landscape (diffuse source pollution), as opposed to more spatially confined sources, such as urban centres and sewage treatment works (point source pollution). But agriculture is also a point source of water pollution, for example, from intensive livestock farms and the disposal of residual pesticides (OECD, 2012).

Policy responses to address agricultural water pollution across OECD countries have typically used a mix of economic incentives, environmental regulations and information instruments. A large range of measures have been deployed at the local, catchment, through to national and transborder scales, across an array of different government agencies. Many approaches to control water pollution from agriculture are voluntary, for example, water supply utilities and the agro-food chain are engaged in co-operative arrangements with farmers to minimise pollution, such as providing farm advisory services. These policies and approaches have had mixed results in lowering agricultural pressure on water systems (OECD, 2012).

Main challenges

The key challenges for policy makers in addressing water quality issues in agriculture are to reduce farm contaminant lost into water systems (negative externalities) while encouraging agriculture to generate or conserve a range of benefits associated with water systems (positive externalities). Clean water is vital in securing economic benefits for agriculture and other sectors, meeting human health needs, maintaining viable ecosystems, and providing societal benefits, such as the recreational, visual amenity, and cultural values society attaches to water systems (OECD, 2012).

9.2. Indicators

Definitions

The indicators related to agricultural water quality include changes in:

- Nitrate, phosphate and pesticide pollution derived from agriculture in surface water, groundwater and marine waters.

Concepts, interpretation, limitations and links to other indicators

Agricultural pollution of water bodies (rivers, lakes, reservoirs, groundwater and marine waters) relates to firstly, the pollution of drinking water, and secondly, the harmful effects on aquatic ecosystems. The latter may result in damage to aquatic organisms, and costs for recreational activities (e.g. swimming), commercial fisheries in both fresh and marine waters, other commercial users (including agriculture) of water drawn downstream, and other values society attaches to water (e.g. visual amenity).

The limitations to identifying trends in water pollution originating from agriculture are in attributing the share of agriculture in total contamination and identifying areas vulnerable to agricultural water pollution. In addition, differences in methods of data collection and national drinking and environmental water standards hinder comparative assessments, while monitoring agricultural water pollution is poorly developed, especially for pesticides, in a number of countries.

The extent of agricultural groundwater pollution is generally less well documented than is the case for surface water, largely due to the costs involved in sampling groundwater, and because most pollutants take a longer time to leach through soils into aquifers. Moreover, the extent of monitoring in agricultural areas in terms of detecting pollution above recommended environmental and recreational use limits is less developed compared to monitoring of drinking water.

A further limitation to the water quality indicators included in this chapter, relate to differences in the monitoring systems used to track nutrient and pesticide pollution of water across countries. These differences include, for example, the number of monitoring stations, the location of monitoring sites in predominantly agricultural water catchments, and the frequency with which readings are taken at a monitoring site, both within a day and at which times during the entire growing season. Hence, comparisons between countries need to be treated cautiously.

The scale of the impairment of water systems due to agriculture described in this chapter also needs to be placed in some perspective. Across most regions in OECD countries drinking water quality is high and there are limited health risks linked to impaired drinking water, although water treatment costs can be significant to remove pollutants (OECD, 2012). Agriculture is also not the only source of contamination of water systems.

Changes in nutrient balances (Chapter 4), pesticide sales (Chapter 5) and soil erosion (Chapter 7) are the key driving forces that are linked to water quality indicators which describe the state of water quality in agricultural areas and define the contribution of nutrient and pesticide pollution originating from agricultural activities (assessment of soil sediment damage from agriculture into water systems is not examined in this report). Adaptation of a range of farm management practices are the response by farmers to reduce pollutant run-off from farmland into water bodies.

Measurability and data quality

Most OECD countries have monitoring networks to measure the overall quality of water systems. However, monitoring of agricultural pollution of water bodies is more limited, with around a half of OECD member countries regularly monitoring nutrient and pesticide pollution (see the figures in this chapter and Annex 1.A2). There are three main sources of information in this chapter, including: the OECD (2008) survey of the overall impacts of agriculture on water systems over the period from 1990 to the mid-2000s across

OECD countries; a more recent OECD (2012) review of national surveys of water quality trends related to agriculture from the mid-2000s to 2010; and updated data provided by OECD countries for this report.

Certain farm pollutants are recorded in more detail and with greater frequency (e.g. nutrients, pesticides), whereas an indication of the overall OECD situation for water pollution from soil sediments, pathogens, salts and other agricultural pollutants is unclear (OECD, 2012). Moreover, pollution levels can vary greatly between OECD countries and regions within countries, depending mainly on soil and crop types, agro-ecological conditions, climate, farm management practices, and policies (Figure 1.1; OECD, 2012).

9.3. Main trends

General overview

The overall trends of agricultural water pollution from nitrates, phosphorus and pesticides across OECD countries are mixed over the period 2000 to 2010, but there appear few situations where significant improvements are reported. Recent national assessments of water pollution related to agriculture, together with limited data on national trends in agricultural water pollution, show a variable picture between countries in terms of the: trends of agricultural water pollution by contaminant type; contribution of agriculture in total pollution; and the extent to which contaminants exceed drinking water standards (OECD, 2012).

For the 15-20 OECD countries that track nutrient and pesticide concentrations in surface water and groundwater, about half record that 10% or more monitoring sites in agricultural areas have concentrations that exceed national drinking water limits (see below Figures 9.5, 9.7 and 9.9). But monitoring sites measuring concentrations in excess of drinking water standards varies greatly between countries, contaminants, and surface and groundwater. For example, the share of monitored sites where pesticide concentrations are above drinking water standards for surface and groundwater supplies are generally lower than for nutrients. But concerns remain for pesticide pollution of groundwater (see below Figures 9.8 and 9.9).

The water consumed by most of the population across OECD countries, however, is well within drinking water standards due to effective treatment to remove these pollutants, which is estimated to cost water treatment companies and consumers billions of dollars annually. But in some rural areas of OECD countries, which are not connected to treated water infrastructure systems, health concerns can be more significant from agricultural water pollution, especially where water is drawn from shallow wells.

The downward trend in nutrient surpluses and pesticide sales over the past 10 years for many OECD countries, however, would suggest that pressure from agriculture on water systems has eased (Chapters 4 and 5). Moreover, overall improvements in slowing rates of soil erosion on agricultural land across many OECD countries, would also indicate that the risk of agricultural water pollution could be declining, as soil sediment is a major pollutant of water systems, including the transportation by soil particles of pollutants into water (Chapter 7; OECD, 2012).

The apparent dichotomy between decreasing agricultural pollutant loads but stable or deteriorating readings of water pollution at monitoring sites, is to a large extent explained by time lags (OECD, 2012). A time lag (sometimes referred to as the legacy problem) is the time elapsed between the adoption of management changes by farmers and the detection of measurable improvement in water quality of the target water body (Fenton et al., 2010; Kronvang, Rubaek and Heckrath, 2009; Meals, Dressing and Davenport, 2010; Schulte et al., 2010).

The magnitude of the time lag is highly site and contaminant specific and can take: hours to months for some contaminants after heavy rainfall, especially point sources in agriculture; years to decades for excessive phosphate levels in agricultural soils; and decades or more for sediment accumulated in river systems (Meals, Dressing and Davenport, 2010). Nutrient enriched lakes and acidified waters may also take years to recover (Environment Agency, 2007). Groundwater travel time is also an important contributor to time lags and may introduce a lag of decades between changes in agricultural practices and improvements in groundwater quality (Collins and McGonigle, 2008; Dubrovsky et al., 2010; Environment Agency, 2007; Meals, Dressing and Davenport, 2010).

Overview for the European Union and other selected OECD countries

Across most **European Union** member states agriculture is an important source of nutrients and pesticides into surface and groundwater (European Environment Agency, 2010). While there are differences in trends and absolute pressures from agricultural nutrient surpluses on water systems across member states, the contribution of agriculture remains high. More specifically in most member states agriculture is responsible for over a third of the total nutrient discharge to surface and coastal waters (Figures 9.1 and 9.3), although the overall trend in agricultural nutrient discharges has been declining since the mid-1990s for some but not all countries (see Figures 9.2, 9.4 and 9.7; and European Environment Agency, 2010; European Commission, 2010). Even so, around a third of EU15 surface water and groundwater monitoring stations still show an upward trend in nitrate concentration levels in water and eutrophication of fresh and marine waters is significant (European Commission, 2010).

Although varying regionally in nature and severity, agricultural influences on water quality are important, according to the **United States** Geological Survey *National Water Quality Assessment Program* (USGS, 2010). Nationally, agriculture is estimated to account for around 60% of river pollution, 30% of lake pollution and 15% of estuarine and coastal pollution. Agriculture also contributes significantly to groundwater contamination (wells and aquifers) across the nation, especially from leaching of nutrients and pesticides (see below Figures 9.6 and 9.8; and USGS, 2010). Of growing concern for groundwater quality is the increasing and widespread detection of contaminant mixtures, including mixtures of pesticides and veterinary products from agriculture with other man-made and natural contaminants (USGS, 2010).

Risks to water quality associated with agriculture currently has a good status in **Canada** (Figure 9.5), but represents an overall decline from a desired state in 1981. Increased application of nutrients (N and P), as fertiliser and manure, has been the main driver for the declining trend in the performance index for agricultural water quality throughout Canada (Eilers et al., 2010). Increased efforts are required across Canada to minimise the risk of nutrient, pesticide and coliform movement to surface water bodies and leaching beyond the rooting depth of vegetation. This is particularly so in higher rainfall areas of the country (Eilers et al., 2010).

Whilst over 90% of rivers meet health related water quality standards in **Japan**, many lakes, reservoirs and coastal waters do not (OECD, 2010). Nitrates, pesticides and sediments from agricultural activities are acknowledged to be among the causes of these problems, as

Figure 9.1. **Agriculture emissions of nitrates and phosphorus in surface water, OECD countries, 2000-09**

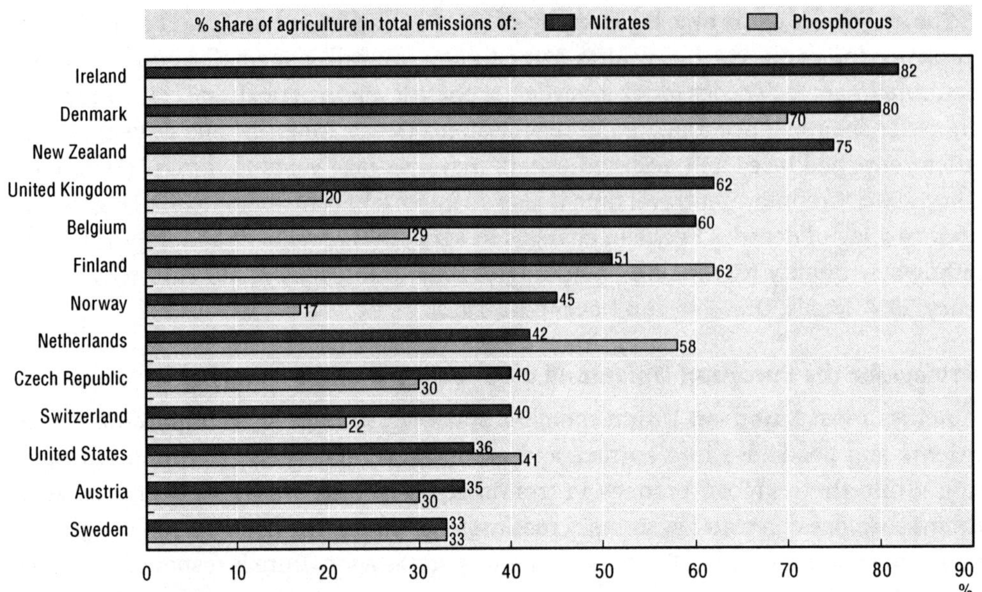

Notes: Countries are ranked in descending order of highest share of nitrates in surface water.

For nitrates, the figures presented correspond to the year 2000 for Austria, Czech Republic, New Zealand, Norway, Switzerland and United States; 2002 for Denmark; 2004 for Finland and Ireland; 2005 for Belgium (Wallonia); 2008 for United Kingdom; and 2009 for Netherlands and Sweden.

For phosphorous, the figures presented correspond to the year 2000 for Austria, Czech Republic, Norway, Switzerland and United States; 2002 for Denmark; 2004 for Finland; 2005 for Belgium (Wallonia); and 2009 for Netherlands, Sweden and United Kingdom.

Source: OECD Agri-Environmental Indicators Questionnaire, unpublished.

StatLink http://dx.doi.org/10.1787/888932793015

Figure 9.2. **Trends in agriculture's emissions of nitrates and phosphorus in surface water, OECD countries, 1995-2009**

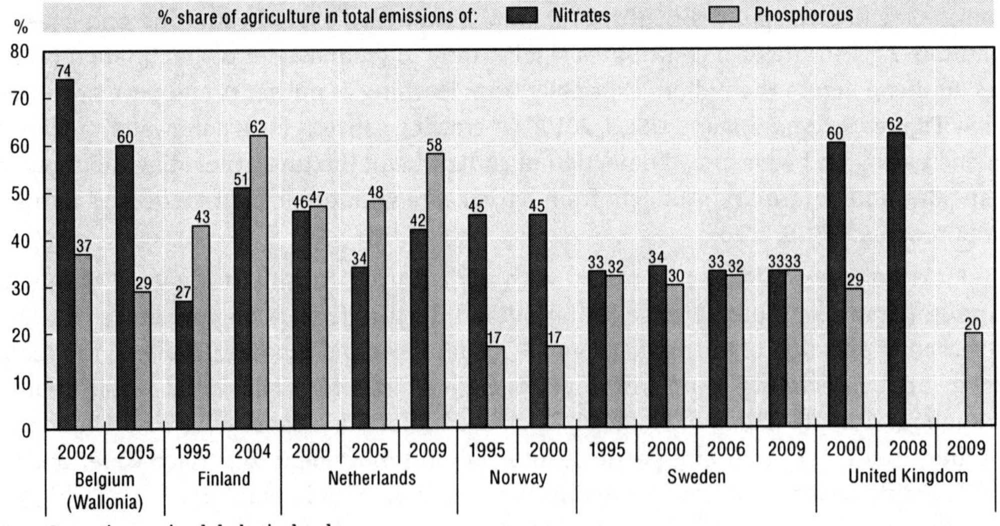

Note: Countries are in alphabetical order.

Source: OECD Agri-Environmental Indicators Questionnaire, unpublished.

StatLink http://dx.doi.org/10.1787/888932793034

Figure 9.3. **Agriculture emissions of nitrates and phosphorus in coastal water, OECD countries, 2000-09**

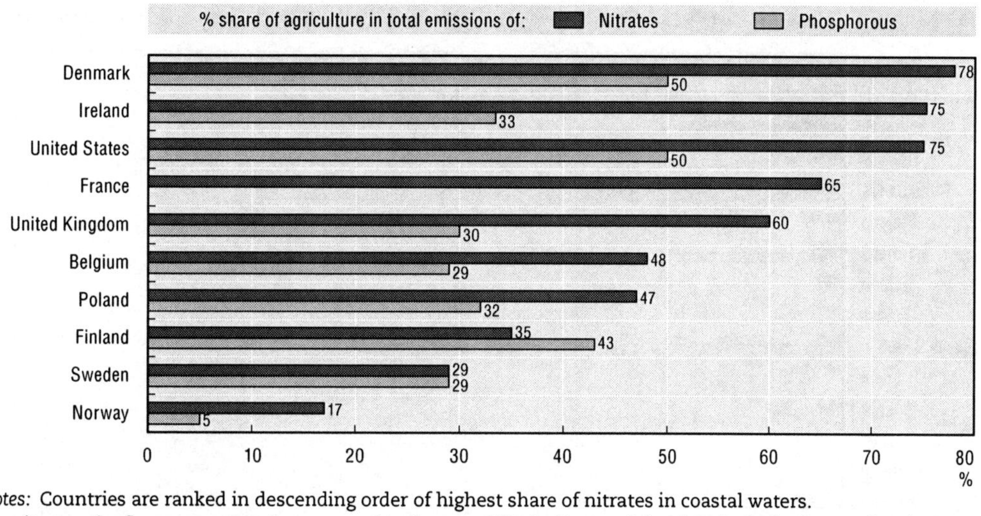

% share of agriculture in total emissions of: Nitrates Phosphorous

Denmark 78 50
Ireland 75 33
United States 75 50
France 65
United Kingdom 60 30
Belgium 48 29
Poland 47 32
Finland 35 43
Sweden 29 29
Norway 17 5

0 10 20 30 40 50 60 70 80 %

Notes: Countries are ranked in descending order of highest share of nitrates in coastal waters.
For nitrates, the figures presented correspond to the year 2000 for France, Poland, United Kingdom and United States; the year 2002 for Denmark; the year 2003 for Belgium (Flanders); the year 2005 for Ireland; the year 2006 for Finland; and the year 2009 for Norway and Sweden.
For phosphorous, the figures presented correspond to the year 2000 for Poland, United Kingdom and United States; the year 2002 for Denmark; the year 2003 for Belgium (Flanders); the year 2005 for Ireland; the year 2006 for Finland; and the year 2009 for Norway and Sweden.
Source: OECD Agri-Environmental Indicators Questionnaire, unpublished.

StatLink http://dx.doi.org/10.1787/888932793053

Figure 9.4. **Trends in agriculture's emissions of nitrates and phosphorus in coastal waters, OECD countries, 1995-2009**

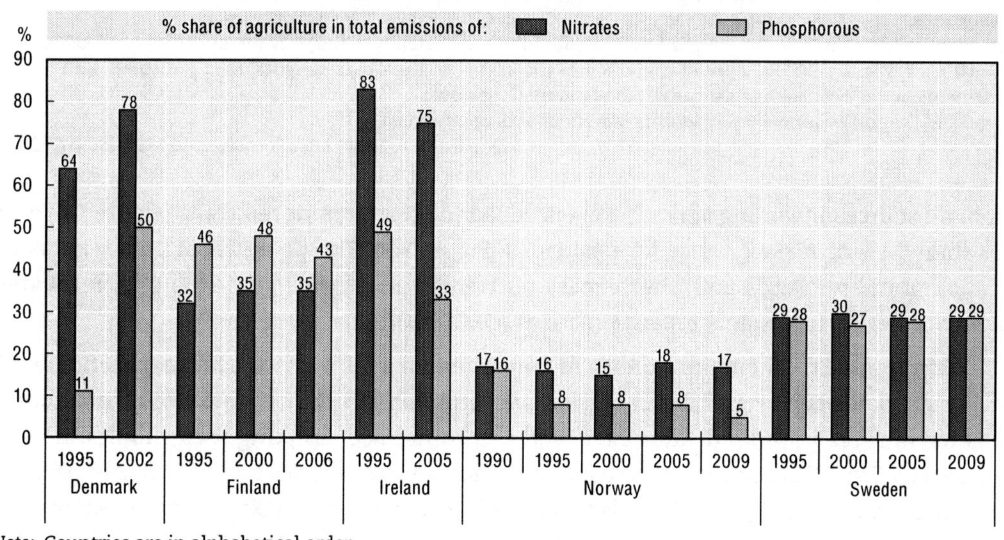

Note: Countries are in alphabetical order.
Source: OECD Agri-Environmental Indicators Questionnaire, unpublished.

StatLink http://dx.doi.org/10.1787/888932793072

well as discharges from other sources (e.g. sewage, industrial). Eutrophication continues to be a concern from nutrients, including from agriculture, especially intensive livestock operations but also fertiliser use, leading to frequent algae blooms (red and blue tides) that damage aquatic life in coastal areas and increase costs of water treatment from inland water intakes (Ileva et al., 2009). The quality of groundwater is improving, with nitrogen

Figure 9.5. **Agricultural areas that exceed recommended drinking water limits for nitrates and phosphorous in surface water, OECD countries, 2000-10**

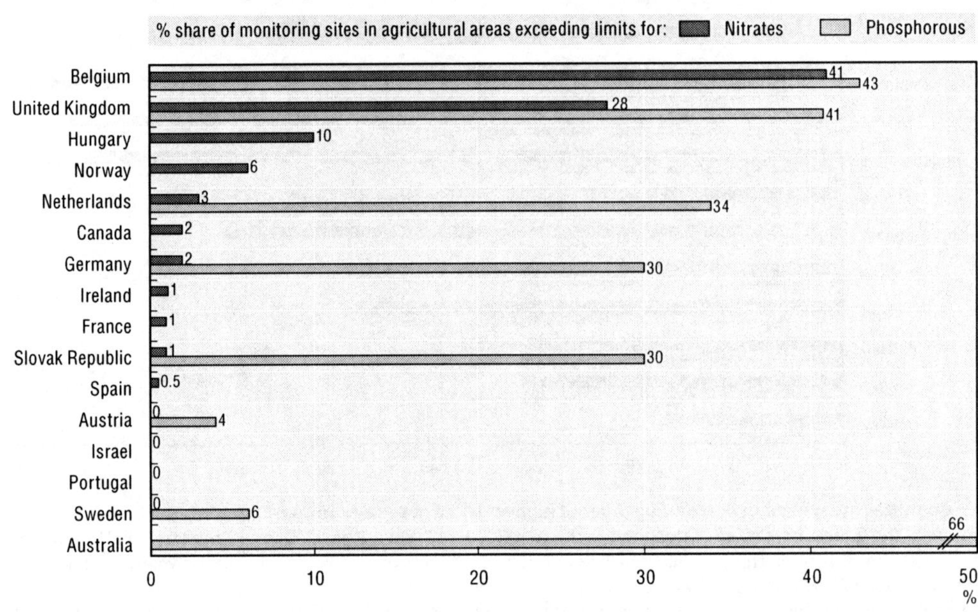

Notes: Countries are ranked in descending order of highest share of monitoring sites exceeding nitrate drinking water limits. For nitrates, data are not available for Australia. For phosphorous, data are not available for Canada, France, Hungary, Ireland, Israel, Norway, Portugal and Spain.

For nitrates, the figures refer to the year 2000 for Austria, Germany, Hungary, Norway and Sweden; the year 2001 for Belgium (Flanders) and Spain; the year 2002 for Slovak Republic; the year 2005 for Portugal; the year 2008 for France; the year 2009 for Canada, Ireland and United Kingdom; and the year 2010 for Israel and Netherlands. Value is zero or less than 0.5% for nitrates for Austria, Israel, Portugal and Sweden.

For phosphorous, the figures refer to the year 2000 for Germany and Sweden; the year 2001 for Austria and Belgium (Flanders); the year 2002 for Australia and Slovak Republic; the year 2009 for United Kingdom; and the year 2010 for Netherlands.

The statistical data for Israel are supplied by and under the responsibility of the relevant Israeli authorities. The use of such data by the OECD is without prejudice to the status of the Golan Heights, East Jerusalem and Israeli settlements in the West Bank under the terms of international law.

Source: OECD Agri-Environmental Indicators Questionnaire, unpublished.

StatLink http://dx.doi.org/10.1787/888932793091

(from all sources including agriculture) exceeded in 5% of monitored wells (Figure 9.6) and less than 0.1% of surface water for pesticides (Figure 9.9). The potential of paddy fields to mimic natural wetlands and filter excess nutrients can provide some benefits for water quality under certain management practices (OECD, 2012).

Despite some recent improvements in **Korea**, around a third of rivers fail to meet domestic quality standards and over a quarter of lakes are eutrophic (OECD, 2012) and 16% of monitoring sites for nitrates in groundwater exceeded recommended drinking water limits in 2000 (Figure 9.6). Coastal eutrophication is a localised problem for fisheries and aquaculture. Diffuse pollution, including from agriculture, is acknowledged as a source of pollution with increases in livestock numbers a growing pressure on water systems (Figure 3.3). Paddy fields mimicking natural wetlands hold the potential to improve water quality.

Problems in **Australia** arising from agricultural contaminants and salinity have been exacerbated by low flow conditions caused by abstraction and less than average rainfall in recent years (OECD, 2012). Most rivers exhibit a high degree of degradation, particularly within the Murray-Darling catchment, Australia's main agricultural producing region. Drinking water quality is impaired in many locations, and coastal regions downstream of large agricultural

Figure 9.6. **Agricultural areas that exceed recommended drinking water limits for nitrates in groundwater, OECD countries, 2000-10**

Country	%
Belgium	32
Denmark	29
Italy	22
Spain	20
United States	20
Korea	16
United Kingdom	16
Greece	15
Slovak Republic	14
France	13
Netherlands	12
Hungary	9
Austria	9
Portugal	6
Japan	5
Switzerland	5
Australia	3
Poland	3
Turkey	3
Finland	2
Ireland	1
Israel	0.4
Norway	0

Notes: Countries are ranked in descending order of highest share of monitoring sites exceeding nitrates limits. The figures refer to 2000 for Japan, Korea, Turkey and United States; 2001 for Greece; 2002 for Australia, Finland, Hungary and Norway; 2003 for Denmark, Italy and Spain; 2005 for Belgium (Flanders), Portugal and Slovak Republic; 2008 for France and Poland; 2009 for Switzerland; and 2010 for Austria, Ireland, Israel, Korea, Netherlands and United Kingdom.
The statistical data for Israel are supplied by and under the responsibility of the relevant Israeli authorities. The use of such data by the OECD is without prejudice to the status of the Golan Heights, East Jerusalem and Israeli settlements in the West Bank under the terms of international law.
Source: OECD Agri-Environmental Indicators Questionnaire, unpublished.

StatLink http://dx.doi.org/10.1787/888932793110

Figure 9.7. **Agricultural areas exceeding national drinking water limits for nitrates in groundwater, Austria, 1990-2010**

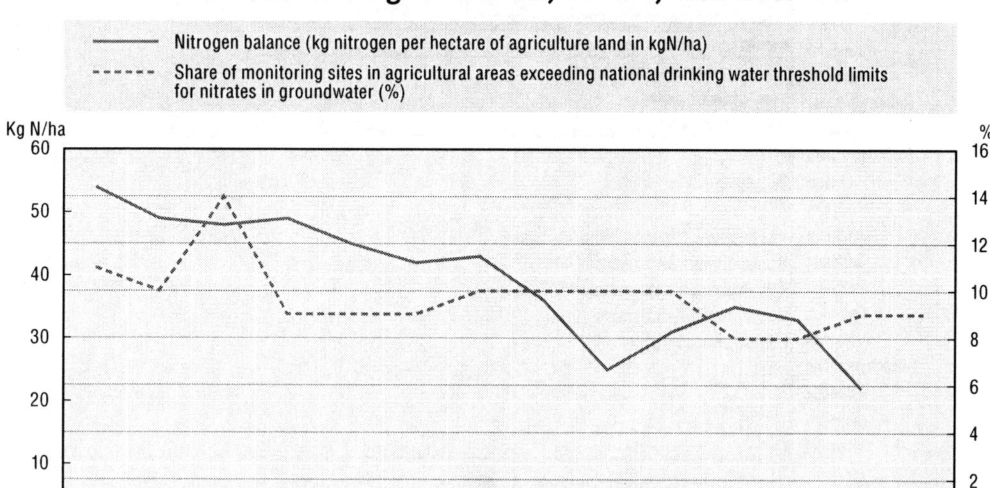

Source: OECD Agri-Environmental Indicators Questionnaire, unpublished.

StatLink http://dx.doi.org/10.1787/888932793129

areas suffer from sediment and nutrient loadings (Figures 9.5, 9.6 and 9.9). In terms of the environmental health of the Great Barrier Reef (GBR), recent research indicates that quantities of sediment, phosphorus and nitrogen entering the GBR have been increasing, with agriculture

Figure 9.8. **Agricultural areas where one or more pesticides are present in surface water and groundwater, OECD countries, 2000-10**

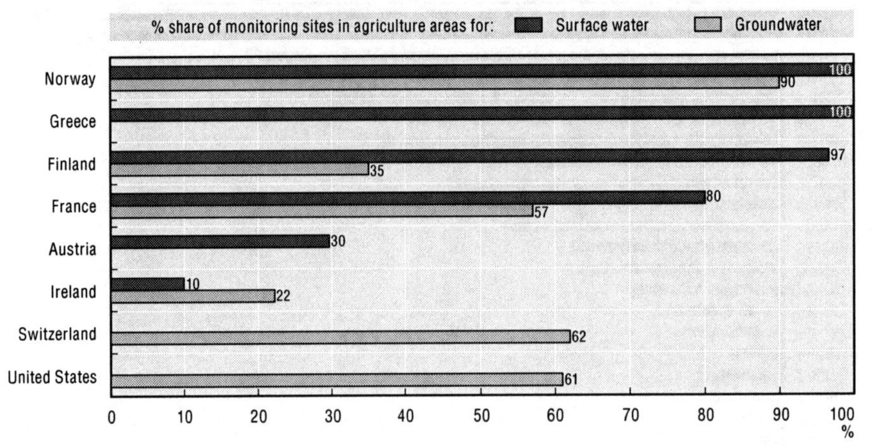

Notes: Countries are ranked in descending order of highest share of pesticide presence in surface water.

Data for surface water are not available for Switzerland and the United States. Data for groundwater are not available for Austria and Greece.

For surface water, the figures presented refer to the 2005-07 average for Austria; the 2008-10 average for Finland and Ireland; the year 2002 for France and Switzerland; the year 2000 for Greece; and the year 2010 for Norway.

For groundwater, the figures presented refer to the 2003-05 average for Finland; the year 2002 for France and United States; the 2007-09 average for Ireland; the year 2009 for Switzerland; and the year 2010 for Norway.

Source: OECD Agri-Environmental Indicators Questionnaire, unpublished.

StatLink http://dx.doi.org/10.1787/888932793148

Figure 9.9. **Agricultural areas where pesticide concentrations in surface water and groundwater exceed recommended national drinking water limits, OECD countries, 2000-10**

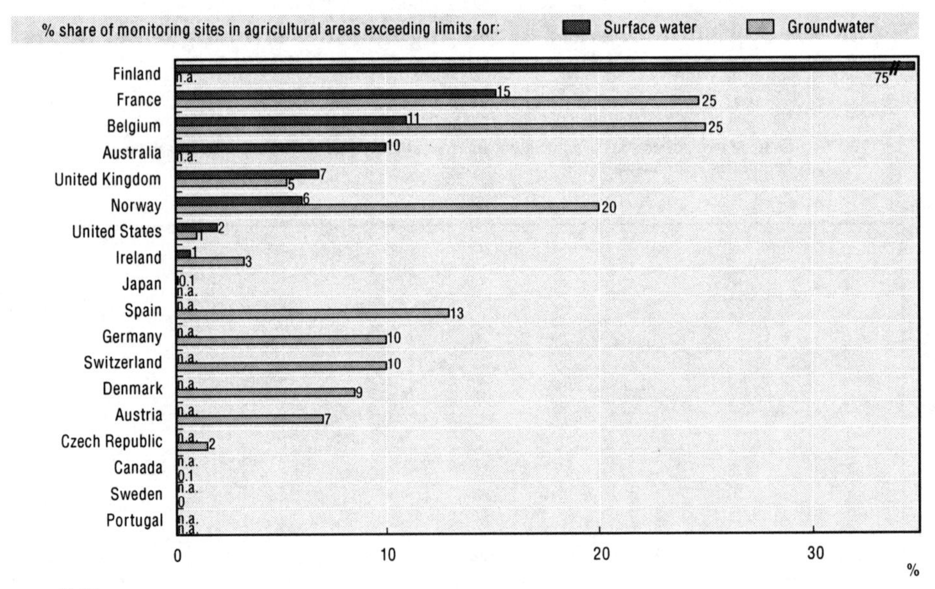

n.a.: not available.

Notes: Countries are ranked in descending order of highest share of monitoring sites exceeding pesticides drinking water limits for surface water.

For surface water, the figures presented refer to: 2002 for Australia, Belgium, France, Norway and United States; 2004 for Ireland; average 1998-2005 for Japan; and average 2005-07 for United Kingdom.

For groundwater, the figures presented refer to: 2000 for Denmark and Germany; 2002 for Belgium, Canada, France, Norway, Spain, Sweden, United Kingdom and United States; 2003 For Czech Republic; average 2007-09 for Ireland; 2009 for Switzerland; and average 2008-10 for Austria.

Source: OECD Agri-Environmental Indicators Questionnaire, unpublished.

StatLink http://dx.doi.org/10.1787/888932793167

a key contributor to water quality issues in the GBR (Rolfe and Windle, 2011). But given the lack of a national monitoring system it is difficult to assess national trends in water quality related to agriculture (State of the Environment 2011 Committee, 2011).

Overall water is of high quality in **New Zealand**, but the quality of a number of lowland rivers and streams is causing concern. There are expensive restoration clean-ups going on in some iconic lakes and there are questions over the state of groundwater. At a national level, diffuse discharges now greatly exceed point source pollution (e.g. sewage treatment works) (Figure 9.1). Around 64% of monitored lakes in pastoral landscapes are classed eutrophic or worse (Ballantine and Davies-Colley, 2009; Land and Water Forum, 2010). Similarly groundwater quality has been deteriorating, with one third of sites monitored between 1995 and 2008 recording increasing nitrate levels (Daughney and Randall, 2009).

Nutrient pollution from nitrates and phosphorus

Agriculture is often the major source of emissions of nitrates and phosphorus into surface water and groundwater across OECD countries (Figures 9.1, 9.2 and 9.6). With point sources of water pollution (i.e. industrial and urban sources) falling more rapidly than for agriculture over several decades, and effectively controlled in most situations, the share of agriculture in nutrient pollution into water systems has been rising even though absolute levels of pollutants have declined in many cases (OECD, 2008; 2012).

Trends in the contribution of nitrate and phosphorus to surface water are mixed over the period 1995 to 2009, however, this conclusion is based on a very limited number of OECD countries (Figure 9.2). For groundwater the information on trends are even more limited, with only **Austria** providing an indication of trends over time (Figure 9.7). In terms of the number of monitoring sites in agricultural dominated areas that exceed recommended drinking water limits for nitrates and phosphorus in surface and groundwater, evidence shows this tends to be lower for surface water (Figure 9.5) compared to groundwater (Figure 9.6).

OECD agriculture is also a significant source of emissions into marine waters (Figures 9.3 and 9.4). Estuarine and coastal agricultural nutrient pollution has been an important contributory factor in some regions leading to eutrophication and the creation of algal blooms (i.e. "red tides" or "dead zones"). This has caused extensive damage to marine life, including commercial fisheries in coastal waters adjacent to **Australia**, **Japan**, **Korea**, the **United States** and **Europe**, mainly the Baltic, North Sea, and Mediterranean (Diaz et al., 2012; OECD, 2012).

In the **Baltic Sea** catchment area, for example, the major anthropogenic source of waterborne nitrogen is diffuse inputs, mainly agriculture (HELCOM 2009; Malmaeus and Karlsson, 2010). They constitute 71% of the total load into surface waters within the catchment area. Agriculture alone contributed about 80% of the reported total diffuse load. The largest loads of phosphorus originated from point sources (56%), with municipalities as the main source, constituting 90% of total point source discharges in 2000, with 44% from diffuse sources, such as agriculture. For some Baltic countries, such as **Finland** and **Sweden**, agriculture is the major contributor of phosphorus into the Baltic.

Similarly the **Gulf of Mexico's** hypoxic (dead/eutrophic) zone, first detected in the 1970s, has increased in size substantially, largely as a result of agricultural nutrients washed from the Mississippi into the Gulf (OECD, 2012). The **United States** National Oceanic and Atmospheric Administration research shows that although the overall trend is for an increase in the area of the Gulf hypoxic zone, the area of the zone varies annually according to climatic conditions (Devine, Dorfman and Rosselot, 2008; Rabotyagov et al., 2010).

Pesticide pollution

The presence of pesticides in surface water and groundwater is widespread across OECD countries, with some countries having over 60% of monitored sites found to have one or more pesticide present in surface water and groundwater (Figure 9.8). But less than a third of OECD countries monitor pesticides in water systems. Caution, however, is required when linking trends in pesticide use to water pollution, as different pesticides pose different types and levels of risks to aquatic environments and drinking water (OECD, 2008).

There are a number of OECD countries where over 10% of monitoring sites in agricultural areas have pesticide concentrations in surface water and groundwater in excess of recommended drinking water limits (Figure 9.9). But as with other agricultural water quality indicators, the number of OECD countries monitoring pesticides in water systems is limited, as are time series data, although data for **Austria** reveals the close link between trends in pesticide sales tracking trends in pesticides detected in surface water, although responses might be delayed for groundwater (Figure 9.10).

Figure 9.10. **Agricultural areas where one or more pesticides are present in surface water and where pesticide concentrations in groundwater exceed recommended national drinking water limits, Austria, 1997-2010**

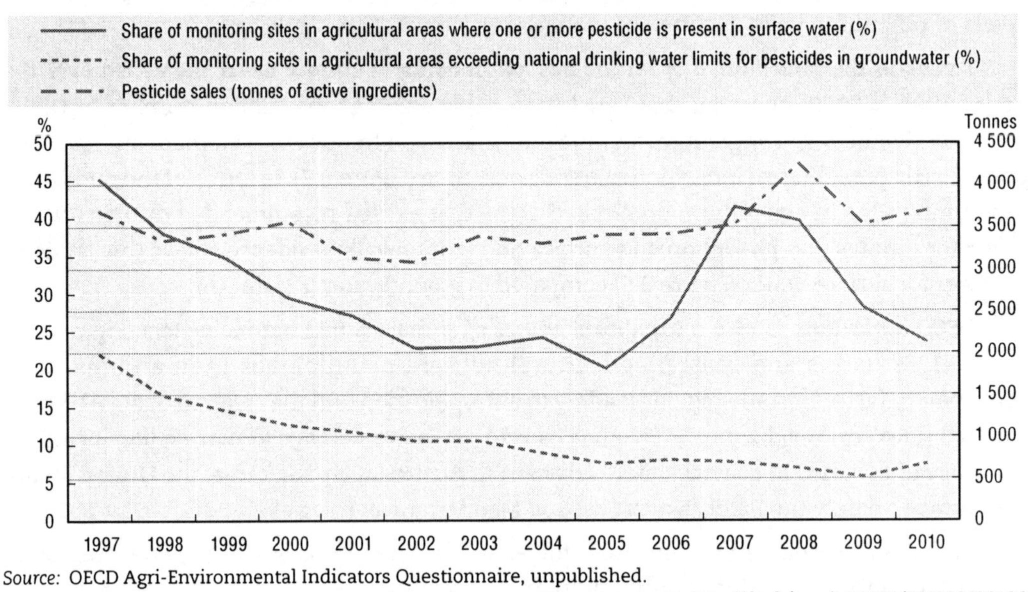

Source: OECD Agri-Environmental Indicators Questionnaire, unpublished.

StatLink http://dx.doi.org/10.1787/888932793186

Another concern with pesticide pollution of water bodies relates to highly persistent and toxic pesticides such as DDT. In most cases in OECD countries such pesticides have been banned for many decades, but are, nevertheless, still being detected at levels harmful to aquatic organisms. This is the case, for example, in **France**, the **United States**, and **Mexico**, although in the latter country the ban on such pesticides was more recent (OECD, 2008). Pesticides are also reported as a common pollutant in coastal waters for some countries (**France** and **Mexico**), with risks to human health from fish consumed from these waters, of particular concern for **Mexico** where pesticide sales have been increasing over the past 20 years (Figure 5.1) (OECD, 2008).

References

Ballantine, D. and R.J. Davies-Colley (2009), *Water Quality Trends at National River Water Quality Network Sites for 1989-2007*, prepared for the New Zealand Ministry for the Environment, National Institute of Water and Atmospheric Research Ltd., Hamilton, New Zealand, *www.mfe.govt.nz/environmental-reporting/freshwater*.

Collins, A.L. and D.F. McGonigle (2008), "Monitoring and Modelling Diffuse Pollution from Agriculture for Policy Support: UK and European Experience", *Environmental Science and Policy*, Vol. 11, pp. 97-101.

Daughney, C. and M. Randall (2009), "National Groundwater Quality Indicators Update: State and Trends 1995-2008", *GNS Consultancy Report* 2009/145, prepared for the New Zealand Ministry of the Environment, Wellington, New Zealand, *www.mfe.govt.nz/environmental-reporting/freshwater*.

Devine, J., M. Dorfman and K.S. Rosselot (2008), "Missing Protection: Polluting the Mississippi River Basin's Small Streams and Wetlands", Natural Resources Defence Council, *Issue Paper*, New York, United States, *www.nrdc.org/water/pollution/msriver/msriver.pdf*.

Díaz, R.J., N.N. Rabalais and D.L. Breitburg (2012), "Agriculture's Impact on Aquaculture: Hypoxia and Eutrophication in Marine Waters", *OECD Consultant Report*, available at: *www.oecd.org/agriculture/water*.

Dubrovsky, N.M. et al. (2010), *The Quality of our Nation's Waters – Nutrients in the Nation's Streams and Groundwater, 1992-2004*, U.S. Geological Survey Circular 1350, U.S. Geological Service, Reston, Virginia, United States.

Eilers, W., R. MacKay, L. Graham and A. Lefebvre (eds.) (2010), "Environmental Sustainability of Canadian Agriculture", *Agri-Environmental Indicator Report Series*, Report No. 3, Agriculture and Agri-Food Canada, Ottawa, Canada, *http://publications.gc.ca/collections/collection_2011/agr/A22-201-2010-eng.pdf*.

Environment Agency (2007), *The Unseen Threat to Water Quality – Diffuse Water Pollution in England and Wales*, Bristol, United Kingdom, May.

European Commission (2010), *Report from the Commission to the Council and the European Parliament on Implementation of Council Directive 91/676/EEC Concerning the Protection of Waters Against Pollution Caused by Nitrates from Agricultural Sources Based on Member State Reports for the Period 2004-2007*, COM(2010)47 Final, Brussels, Belgium.

European Environment Agency (2010), *The European Environment State and Outlook 2010: Freshwater Quality*, Copenhagen, Denmark, *www.eea.europea.eu*.

Fenton, O., R.O. Schulte, P. Jordan, S.T.J. Lalor and K. Richards (2010), "Lag Time: A Methodology for the Estimation of Vertical and Horizontal Travel and Flushing Timescales to Nitrate Threshold Concentrations in Irish Aquifers", *Environmental Science and Policy*, Vol. 14, pp. 419-431.

HELCOM (Helsinki Commission) (2009), "Eutrophication in the Baltic Sea: An Integrated Thematic Assessment of the Effects of Nutrient Enrichment in the Baltic Sea Region – Executive Summary", *Baltic Sea Environment Proceedings*, No. 115A, HELCOM, Helsinki, Finland.

Ileva, N.Y., H. Shibata, F. Satoh, K. Sasa and H. Ueda (2009), "Relationship Between the Riverine Nitrate-Nitrogen Concentration and the Land Use in the Teshio River Watershed, North Japan", *Sustainability Science*, Vol. 4, pp. 189-198.

Kronvang, B., G.H. Rubaek and G. Heckrath (2009), "International Phosphorus Workshop: Diffuse Phosphorus Loss to Surface Water Bodies – Risk Assessment, Mitigation Options, and Ecological Effects in River Basins", *Journal of Environmental Quality*, Vol. 38, pp. 1924-1929.

Land and Water Forum (2010), *Report of the Land and Water Forum: A Fresh Start for Freshwater*, Wellington, New Zealand, *www.landandwater.org.nz*.

Malmaeus, J.M. and O.M. Karlsson (2010), "Estimating Costs and Potentials of Different Methods to Reduce the Swedish Phosphorus Load from Agriculture to Surface Water", *Science of the Total Environment*, Vol. 408, pp. 473-479.

Meals, D.W., S.A. Dressing and T.E. Davenport (2010), "Lag Time in Water Quality Response to Best Management Practices: A Review", *Journal of Environmental Quality*, Vol. 39, pp. 85-96.

OECD (2012), *Water Quality and Agriculture: Meeting the Policy Challenge*, OECD Publishing, *www.oecd.org/agriculture/water*.

OECD (2010), *Environmental Performance Reviews: Japan*, OECD Publishing, *www.oecd.org/env*.

OECD (2008), *Environmental Performance of Agriculture in OECD Countries Since 1990*, OECD Publishing, *www.oecd.org/tad/sustainable-agriculture/agri-environmentalindicators.htm*.

Rabotyagov, S., T. Campbell, M. Jha, P.W. Gassman, J. Arnold, L. Kurkalova, S. Secchi, H. Feng and C.L. Kling (2010), "Least-Cost Control of Agricultural Nutrient Contributions to the Gulf of Mexico Hypoxic Zone", *Ecological Applications*, Vol. 20, No. 6, pp. 1542-1555.

Rolfe, J. and J. Windle (2011), "Using Auction Mechanisms to Reveal Costs for Water Quality Improvements in Great Barrier Reef Catchments in Australia", *Agricultural Water Management*, Vol. 98, pp. 493-501.

Schulte, R.P.O., A.R. Melland, O. Fenton, M. Herlihy, K. Richards and P. Jordan (2010), "Modelling Soil Phosphorus Decline: Expectations of Water Framework Directive Policies", *Environmental Science and Policy*, Vol. 13, pp. 472-484.

State of the Environment 2011 Committee (2011), *Australia State of the Environment 2011*, Department of Sustainability, Environment, Water, Population and Communities, Australian Government, Canberra, Australia.

USGS (U.S. Geological Survey) (2010), *Quality of Water from Public-Supply Wells in the United States, 1993-2007: Overview of Major Findings*, US Geological Survey, Circular 1346, Washington, DC, United States, *pubs.usgs.gov/sir/2010/5024*.

Chapter 10

Ammonia emissions: Acidification and eutrophication

This chapter reviews the environmental performances of agriculture in OECD countries related to ammonia emissions. It provides a description of the policy context (issues and main challenges), definitions for the agri-environmental indicators presented, and elements related to concepts, interpretations, links to other indicators, as well as measurability and data quality. The chapter then describes the main trends of the agri-environmental indicators, using available data covering the period 1990-2010 and based on a set of tables and figures.

This chapter, together with the following Chapter 11 on greenhouse gases, examines how agricultural activities impact on air quality, through emissions of ammonia (NH_3) and greenhouse gases (methane CH_4, nitrous oxide N_2O and carbon dioxide CO_2). The environmental impacts of these agricultural emissions should be viewed in the broader context of other pollution sources (e.g. industry, transport) and considered in terms of the chemical reactions between different air pollutants in the atmosphere ("multi-pollutants", e.g. sulphur dioxide, carbon dioxide) and the resultant effects on the environment ("multi-effects", e.g. acidification, eutrophication) (Figure 10.1).

Figure 10.1. **Impacts of agriculture on air quality: Multi-pollutants, multi-effects**

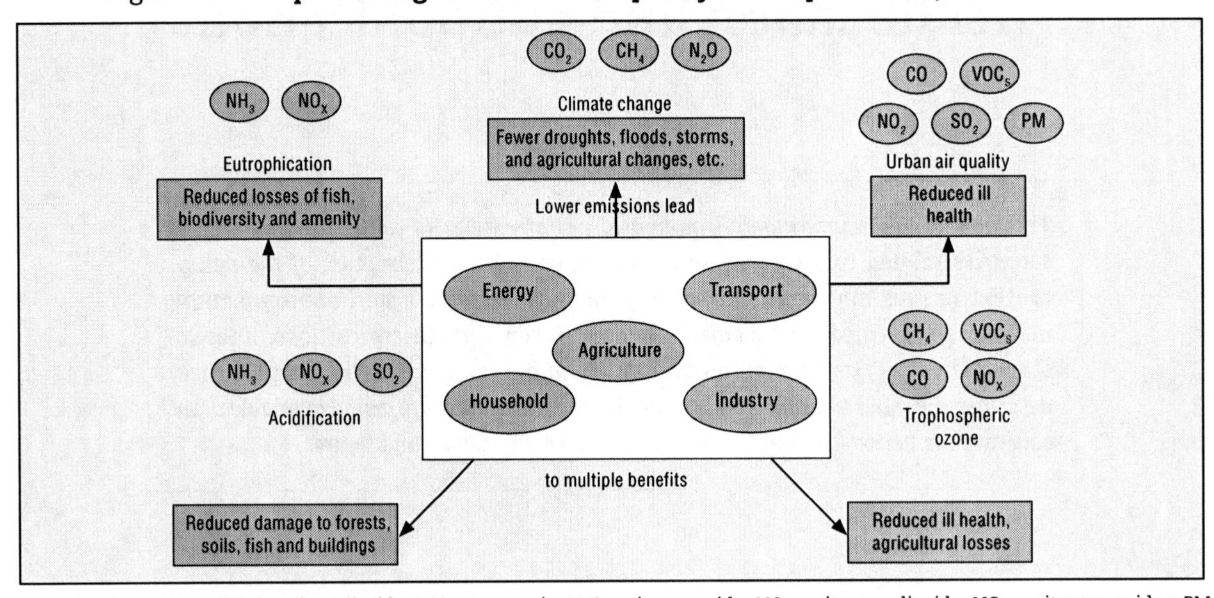

Notes: CH_4 – methane; CO_2 – carbon dioxide; NH_3 – ammonia; N_2O – nitrous oxide; NO_2 – nitrogen dioxide; NO_X – nitrogen oxides; PM – particulate matter; SO_2 – sulphur dioxide; VOCs – volatile organic compounds.
Source: EEA (2000), "Environmental Signals 2000", *Environmental Assessment Report*, No. 6, Copenhagen, Denmark.

10.1. Policy context

The issue

Ammonia emissions can have potential adverse impacts on human and animal health and on the environment. Close to the source of emission, high concentrations of ammonia may affect the respiratory system of human beings and animals, and disrupt the physiology of plants, and contribute, at longer distances from the source, to the acidification and the eutrophication of soils and water. Moreover, ammonia is a precursor to fine particulate matter (PM 2.5). In some conditions ammonia will combine with other air-borne chemicals and particles to become fine particulate matter that triggers health problems such as asthma, lung cancer, cardiovascular issues, birth defects, and premature death.

Due to its adverse impacts on health and ecosystems, many OECD countries have made efforts to reduce ammonia emissions from the agricultural sector. In addition, effort has been undertaken to develop technologies and management practices to reduce the level of ammonia emissions, particularly relating to how manure is managed from storage through to spreading. The uptake of these technologies and practices has been encouraged through: regulatory measures concerning the storage of livestock manure and spreading; government farm extension services; and financial support provided to farmers, such as for improved manure storage facilities (OECD, 2008; 2012).

As an atmospheric gas ammonia is very mobile and can move across national boundaries. In an international effort to curb ammonia and other acidifying emissions the Protocol to Abate Acidification, Eutrophication and Ground Level Ozone (the *Gothenburg Protocol*) was adopted in 1999 by some OECD countries under the *Convention on Long-range Transboundary Air Pollution* (LRTAP Convention). The *Gothenburg Protocol* set national ammonia emission ceilings for 2010, except for **Canada** and the **United States**, while the **EU** Directive on *National Emission Ceilings* (NECD) (2001/81/EC) set ammonia emission ceilings at levels identical to those of the *Gothenburg Protocol*.

A proposal for a revised *Gothenburg Protocol* was under international negotiation in 2012, which is expected to include emission ceilings to be met by 2020 for ammonia and other air emissions under the LRTAP. At the same time the **EU's** NECD is also being reviewed, as part of the implementation of the EU's *Thematic Strategy on Air Pollution*, and a proposal for a revised NECD is expected by 2013 (European Environment Agency, 2012; UNECE, 2012).

Main challenges

The key challenge for ammonia emissions from agriculture is to reduce total emissions to: remove its toxic effects on human and animal health, especially inside or close to farm buildings; reduce ecological degradation within ecosystems that are sensitive to an oversupply of nitrogen; diminish its role in impacting climate change as ammonia emissions also lead to the release of nitrous oxide (N_2O), a greenhouse gas; and minimise the economic loss of an essential plant nutrient. There are several means of reducing the adverse impacts of ammonia emissions, but these largely focus on livestock and inorganic fertilisers the main sources of ammonia emissions from agriculture. Most losses of agricultural ammonia emissions derive from livestock housing and grazing and applying manure to fields.

10.2. Indicators

Definitions

The indicator for agricultural ammonia (NH_3) emissions is defined as the change in:

- Agricultural ammonia emissions.

Concepts, interpretation, limitations and links to other indicators

Ammonia (NH_3) emissions are associated, as a driving force, with two major types of environmental issues: acidification and eutrophication (Figure 10.1). Ammonia along with sulphur dioxide (SO_2) and nitrogen oxides (NO_X) contribute to *acidification* of soil and water when it combines with water in the atmosphere or after deposition (OECD, 2008). Excess soil acidity may be damaging to certain types of terrestrial and aquatic ecosystems. As a source of nitrogen, deposition of ammonia can also raise nitrogen levels in soil and water, which may contribute to *eutrophication* in receiving aquatic ecosystems. Along with acidification and eutrophication, agricultural NH_3 emissions may be a significant contributor to the formation of aerosols in the atmosphere which may impair human health (i.e. worsen respiratory conditions), visibility, climate, and produce an unpleasant odour close to the farm (Eilers et al., 2010).

The indicator of agricultural ammonia emissions requires careful interpretation due to a number of limitations in its calculation (Eilers et al., 2010). The key underlying information to calculate agricultural ammonia emissions could be improved in many cases, including information on livestock feeding, housing and manure storage and spreading practices. Moreover, while standardised national emission factors are used for some calculations, emissions are known to vary through the year and also emissions can have spatially varying impacts on human health and the environment.

The agricultural ammonia emission indicator is linked to trends in nitrogen balances (Chapter 4) and greenhouse gas emissions (Chapter 11) as driving forces on the state or concentrations of nitrates in water bodies (Chapter 9) and acidifying pollutants in the air. The agriculture sector in many OECD countries is obliged to respond by reaching national ammonia emission ceilings agreed under the *Gothenburg Protocol* (see Table 10.2 below), through for example the adoption of nutrient management practices.

Measurability and data quality

For those countries under the *Gothenburg Protocol*, the ammonia emission data used in this chapter are drawn from EMEP, the European Environment Agency (EEA, 2012) and national sources for other countries (Figure 10.2). As part of the effort to reduce ammonia emissions from agriculture, in many OECD countries considerable research has been undertaken to validate and improve the emission factors that are used in estimating the level of ammonia emissions (UNECE/EMEP, 2009). While data quality related to ammonia emissions is high, despite the limitations noted above, a key problem is that data series are not available for almost a third of OECD countries.

10.3. Main trends

Overall trends in OECD agricultural ammonia emissions declined at -1.3% per annum between 1998-2000 to 2008-10, following a small increase over the 1990s (Figure 10.2). This conclusion needs to be qualified because a fifth of OECD countries do not report ammonia emission trends. **Canada**, **Finland** and **Israel** all recorded an increase in ammonia emissions. For all three countries livestock production, the main source of agricultural ammonia emissions, increased over the 2000s. It is likely, however, that emissions also rose for those OECD countries where data are not available, especially **Chile**, **Iceland**, **Korea**, **Mexico** and **New Zealand**, given the increases in livestock production in these countries over the last decade, the major source of ammonia emissions for most countries, as discussed further below (Figure 3.3).

A notable trend in ammonia emissions over the past 20 years, has been the reduction in emissions for all the EU transition countries (**Czech** and **Slovak Republics, Estonia, Hungary, Poland** and **Slovenia**). This can be mostly explained by the sharp contraction of the agricultural sectors of these countries following their transition to a market economy during the 1990s. In turn, this was followed for these countries by a period of some

Figure 10.2. **Ammonia emissions from agriculture, OECD countries, 1990-2010**

□ 1990-92 to 1998-2000
■ 1998-2000 to 2008-10

-9.5% -7% -6.6% -8.3%

-5 % 0 5

	Average (thousand tonnes)			Average annual % change		Total ammonia emissions 2008-10[4] from all sources (thousand tonnes)	Agricultural % share in total ammonia emissions 2008-10
	1990-92[1]	1998-2000[2]	2008-10[3]	1990-92 to 1998-2000	1998-2000 to 2008-10		
Israel	..	12	15	..	3.1	19	82
Canada	366	414	441	1.6	0.7	492	90
Finland	34	33	34	-0.2	0.2	37	90
Spain	295	335	333	1.6	-0.1	362	92
Estonia	22	10	10	-9.5	-0.1	10	93
Switzerland	66	59	58	-1.4	-0.1	63	92
Norway	20	21	20	0.6	-0.2	23	90
Luxembourg	5	5	4	-0.6	-0.5	5	94
Turkey	1	1	1	-0.1	-0.6	1	98
Slovenia	19	18	17	-1.0	-0.6	18	95
Austria	64	62	58	-0.3	-0.6	63	93
Hungary	98	70	66	-4.1	-0.7	67	97
Italy	424	414	372	-0.3	-1.1	394	94
Germany	644	594	528	-1.0	-1.2	564	94
France	738	732	640	-0.1	-1.3	658	97
OECD[5]	**7 981**	**8 049**	**7 040**	**0.1**	**-1.3**	**7 670**	**92**
EU15	**3 379**	**3 119**	**2 720**	**-1.0**	**-1.4**	**2 908**	**94**
Ireland	113	121	106	0.9	-1.4	107	98
Greece	78	71	61	-1.1	-1.5	64	96
Czech Republic	131	74	63	-7.0	-1.5	66	95
Sweden	49	51	43	0.6	-1.7	51	84
Poland	406	324	271	-2.8	-1.8	277	98
Portugal	52	53	45	0.1	-1.8	51	88
United States[6]	3 421	3 910	3 349	1.7	-1.9	3 721	90
United Kingdom	342	309	251	-1.3	-2.1	283	89
Slovak Republic	53	31	24	-6.6	-2.3	25	97
Belgium	105	83	64	-3.0	-2.6	69	93
Denmark	123	100	73	-2.6	-3.1	76	96
Netherlands	314	157	108	-8.3	-3.7	126	86
Korea	143	181	..	2.4	..	..	..

Notes: Countries are ranked in terms of highest to lowest average annual % change for 1998-2000 to 2008-10.
The statistical data for Israel are supplied by and under the responsibility of the relevant Israeli authorities. The use of such data by the OECD is without prejudice to the status of the Golan Heights, East Jerusalem and Israeli settlements in the West Bank under the terms of international law.

1. Data for 1990-92 average equal to the year 1990 for Korea.
2. Data for 1998-2000 average equal to the 2000-02 average for Israel.
3. Agricultural ammonia emissions, data for 2008-10 average equal to the 2007-09 average for Canada, Israel, Portugal and the average 2006-08 for United States.
4. For total ammonia emissions, data for 2008-10 average equal to the year 2009 for Israel and 2010 for Turkey.
5. Due to unavailable data, OECD does not include Australia, Chile, Iceland, Israel (1990-99), Japan, Korea, Mexico and New Zealand.
6. Data for agricultural ammonia emissions have been estimated based on the ratio agricultural ammonia/total ammonia emissions using the share 90% as recommended by USEPA.

Source: EMEP (2012), website of the Co-Operative Programme for Monitoring and Evaluation of the Long-Range Transmission of Air Pollutants in Europe (EMEP), *http://webdab1.umweltbundesamt.at/scaled_country_year.html* and national sources for Finland, Israel, Italy, Netherlands, Norway, Poland, Portugal, Slovenia, Switzerland and Turkey.

StatLink *http://dx.doi.org/10.1787/888932793205*

agricultural recovery over the 2000s, when the rate of emission release, although continuing to decline, was at a slower rate than over the 1990s (Figure 10.2).

National trends in agricultural NH_3 emissions mask important regional variations within countries. For example, emissions are highest in Northern **Italy** because of the more intense use of fertilisers in the region, and in **France**, Brittany has the highest emission levels because of the concentration of intensive livestock production in the region (OECD, 2008). Also ammonia emissions and acidification of soils and acidifying precipitation shows considerable regional variation across **Canada** (Eilers et al., 2010) and the **United States** (National Atmospheric Deposition Program).

Agriculture is the main source of ammonia emissions in OECD member countries, accounting for 92% of emissions in 2008-10, ranging from 82-98% (Figure 10.2). The total OECD emissions of acidifying gases (SO_2, NO_X and NH_3) are declining, mainly due to a substantial reduction in SO_2 and NO_X emissions from industry and the energy sector, compared to the more modest reduction for ammonia (Table 10.1). As a consequence the share of agricultural ammonia emissions in total acidifying emissions has risen, despite the absolute reduction in agricultural ammonia emissions over the 2000s (Table 10.1).

Table 10.1. **Total emissions of acidifying pollutants, OECD countries, 1990-2010**

	1990-92		2008-10		1990-92 to 2008-10	
	Average	Share of total	Average	Share of total	Total change	
	Thousand tonnes acid equivalents[1]	%	Thousand tonnes acid equivalents[1]	%	Thousand tonnes acid equivalents[1]	% change
Sulphur dioxide (SO_2)[2]	1 486	51	488	33	-998	-67
Nitrogen oxides (NO_X)[3]	927	32	551	37	-377	-41
Ammonia (NH_3)[4]	515	18	450	30	-65	-13
Total	**2 928**	**100**	**1 488**	**100**	**-1 440**	**-120**

1. The following weighting factors are used to combine emissions in terms of their potential acidifying effect acid equivalent/g: SO_2 = 1/32, NO_X = 1/46 and NH_3 = 1/17.
2. OECD total for SO_2 excludes Australia, Chile (2007-10), Israel (1990-95, 1997-99, 2001-02), Japan, Korea, Mexico (1990-98, 2000-04, 2006-10), New Zealand and Turkey (1990-2009). Portugal excludes Land-Use, Land-Use Change and Forestry (LULUCF). Data for 2010 equal to the 2007-09 average for Israel, Portugal and United States.
3. OECD total for NO_2 excludes Australia, Chile (2007-10), Iceland, Israel (1990-95, 1997-99, 2001-02), Japan, Korea (2008-10), Mexico (1990-98, 2000-04, 2006-10), New Zealand and Turkey (1990-2009). Portugal excludes LULUCF. Data for 2010 equal to the 2007-09 average for Israel, Portugal and United States.
4. OECD total NH_3 excludes Australia, Chile (1990-2004), Iceland, Israel (1990-08), Japan, Korea, Mexico (1990-98, 2000-10), New Zealand and Turkey (1990-09).

Source: EMEP (2012), website of the Co-operative Programme for Monitoring and Evaluation of the Long-Range Transmission of Air Pollutants in Europe (EMEP), *http://webdab1.umweltbundesamt.at/scaled_country_year.html*; European Environment Agency and national sources for Finland, Israel, Italy, Netherlands, Norway, Poland, Portugal, Switzerland and Turkey.

StatLink *http://dx.doi.org/10.1787/888932793490*

Because ammonia is highly reactive, high concentrations (enough to cause odours and significant nitrogen deposition) usually occur close (i.e. less than 2 km) to the emission source (OECD, 2008). In terms of the deposition of NH_3 while around 20% of emissions are deposited close to the source, the rest can travel long distances through the atmosphere, including travelling across national boundaries, which is particularly important for European OECD countries.

The reduction in agricultural ammonia emissions can be expected to have reduced the pressure on the health of humans and animals, as well as ecosystems. It is outside the scope of this report to provide a systematic review of the data and literature concerning the impacts of agricultural ammonia emissions on health and the environment. However, in undertaking such a review there are a number of difficulties, including separating out the presence of other acidifying pollutants, unravelling the spatial and temporal aspects of emissions, and the continued lack of knowledge and data of these impacts (OECD, 2008).

Agricultural ammonia emissions mainly derive from livestock (manure and slurry), and to a lesser extent from the application of inorganic fertilisers to crops and also from

Figure 10.3. **Agricultural ammonia emissions and livestock production volume, OECD countries, 1990-2010**

Panel A. Average annual % change between 1990-92[1] and 1998-00[2]

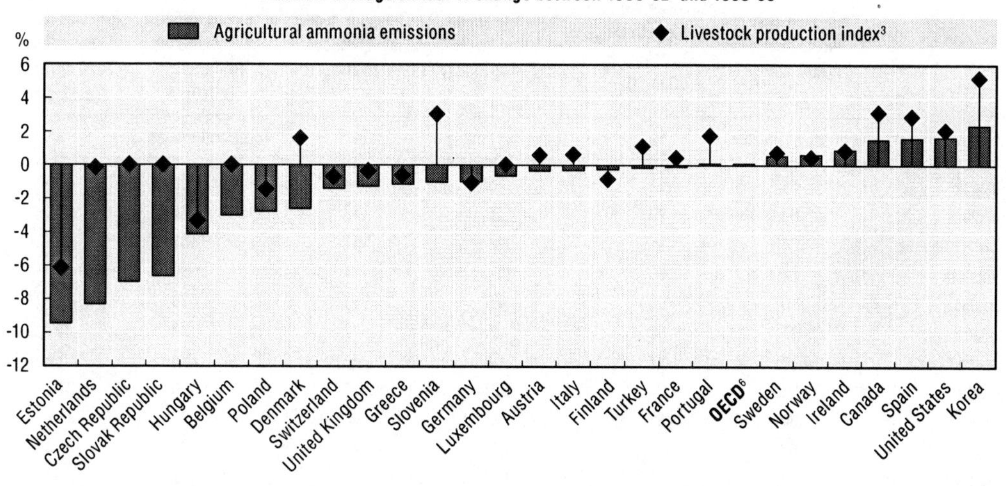

Panel B. Average annual % change between 1998-00 and 2008-10[4]

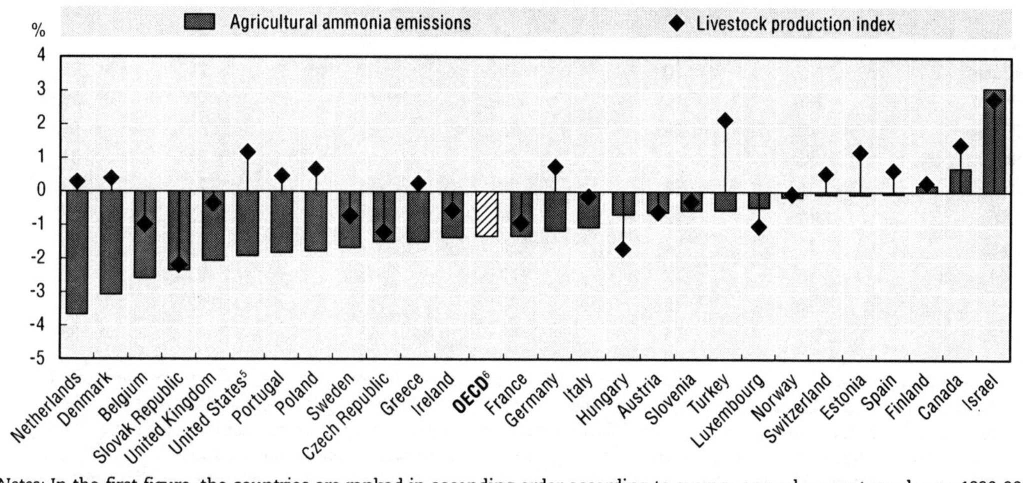

Notes: In the first figure, the countries are ranked in ascending order according to average annual percentage change 1990-92 to 1998-2000 for ammonia emissions. In the second figure, the countries are ranked in ascending order according to average annual percentage change 1998-2000 to 2008-10 for ammonia emissions.

The statistical data for Israel are supplied by and under the responsibility of the relevant Israeli authorities. The use of such data by the OECD is without prejudice to the status of the Golan Heights, East Jerusalem and Israeli settlements in the West Bank under the terms of international law.

1. For livestock production indices, data for 1990-92 average equal to the year 1990 for Korea and the year 1992 for Estonia and Slovenia. For agricultural ammonia emissions, the 1990-92 average refers to the year 1990 for Korea.
2. For livestock production indices, data for 1998-2000 average equal to the year 2000 for Belgium and Luxembourg, 2000-02 average for Israel.
3. The FAO indices of agricultural livestock production show the relative level of the aggregate volume of agricultural production for each year in comparison with the base period 2004-06. They are based on the sum of price weighted quantities of different livestock commodities. All the indices at the country, regional and world levels are calculated by the Laspeyres formula. Production quantities of each commodity are weighted by 2004-06 average international commodity prices and summed for each year. To obtain the index, the aggregate for a given year is divided by the average aggregate for the base period 2004-06. Due to technical reasons it is not possible to provide an OECD or EU average.
4. For livestock production indices and agricultural ammonia emissions, data for 2008-10 average equal to the 2007-09 average for Canada, Israel, Portugal and the average 2006-08 for United States.
5. Data for agricultural ammonia emissions have been estimated based on the ratio agricultural ammonia/total ammonia emissions using the share 90% as recommended by USEPA.
6. Due to unavailable data for agricultural ammonia emission, OECD does not include Australia, Chile, Iceland, Israel (1990-99), Japan, Korea, Mexico, New Zealand.

Source: EMEP (2012), website of the Co-operative Programme for Monitoring and Evaluation of the Long-Range Transmission of Air Pollutants in Europe (EMEP), *http://webdab1.umweltbundesamt.at/scaled_country_year.html*; and national sources for Finland, Israel, Italy, Netherlands, Norway, Poland, Portugal, Switzerland and Turkey; FAOSTAT (2012), *http://faostat.fao.org*.

StatLink http://dx.doi.org/10.1787/888932793224

Table 10.2. Ammonia emission 2010 ceilings under the Convention on Long-Range Transboundary Air Pollution, OECD countries, 1990-2010[1]

	Total ammonia emissions (2008-10)	Share of agriculture emissions in total emissions (2008-10)	Total ammonia emission levels (1990 – base year)	Change in total emission reductions (1990 to 2008-10)	Total emission ceilings under the Gothenburg Protocol[1] (2010)	Total emissions for 2008-10 as a share of the 2010 emission ceilings[2]
	Thousand tonnes	%	Thousand tonnes	%	Thousand tonnes	%
Austria	63	93	65	-4	66	95
Belgium	69	93	120	-43	74	93
Czech Republic	66	95	156	-57	101	66
Denmark	76	96	114	-34	69	110
Estonia	10	93	25	-58	28	37
Finland	37	90	38	-2	31	121
France	658	97	704	-6	780	84
Germany	564	94	692	-18	550	103
Greece	64	96	85	-25	73	87
Hungary	67	97	124	-46	90	75
Ireland	107	98	107	1	116	93
Italy	394	94	468	-16	419	94
Luxembourg	5	94	5	-13	7	68
Netherlands	126	86	355	-65	128	98
Norway	23	90	21	9	23	99
Poland	277	98	508	-46	468	59
Portugal[3]	49	88	63	-22	108 (90)[4]	46
Slovak Republic	25	97	65	-62	39	64
Slovenia	18	95	21	-16	20	89
Spain	362	92	318	14	353	102
Sweden	51	84	55	-7	57	90
Switzerland	63	92	73	-13	63	101
United Kingdom	283	89	360	-21	297	95
EU15	2 907	94	3 549	-18	3 128	93

1. The following countries are not signatories to the Gothenburg Protocol of the Convention on Long-Range Transboundary Air Pollution: Australia, Canada, Iceland, Israel, Korea, Japan, Mexico, New Zealand and Turkey. Canada and the United States are signatories to the LRTAP, but have no emission targets, so are not included in the table.
2. This column shows, for each respective country, the extent to which emissions in 2008-10 were below the emission ceilings for 2010 (e.g. Czech Republic) or exceed the emission ceiling (e.g. Denmark), by dividing the total emission ceiling for 2010 by the total emissions for 2008-10.
3. For total ammonia emissions and total agricultural ammonia emissions, data for 2008-10 average refer to 2007-09 for Portugal.
4. The figure in brackets for Portugal refers to the emission ceiling under the EU Directive on National Emission Ceilings for Certain Atmospheric Pollutants, October 2001 (European Communities, 2001), *http://europa.eu.*

Source: EMEP (2012), website of the Co-operative Programme for Monitoring and Evaluation of the Long-Range Transmission of Air Pollutants in Europe (EMEP), *http://webdab1.umweltbundesamt.at/scaled_country_year.html*; and national sources for Finland, Italy, Netherlands, Norway, Poland, Portugal, Slovenia and Switzerland.

StatLink http://dx.doi.org/10.1787/888932793509

decaying crop residues. For most OECD countries over 90% of total NH_3 emissions are derived from livestock, but for a few countries (e.g. **Korea**, **Japan**) the share of emissions from fertiliser use is over 20%, reflecting the greater importance of the crop sector in these countries relative to those where livestock dominant (OECD, 2008).

The strong link between ammonia emissions and changes in livestock production is highlighted in Figure 10.3, although this is not a linear relationship. Changes in ammonia emissions are positively linked to changes in livestock, illustrating the importance of this driver for the majority of OECD countries, both for countries where livestock production has been increasing (e.g. **Canada**) or decreasing (e.g. **Belgium**). For most, but not all, countries the decrease in agricultural ammonia emissions are always greater than the decrease in livestock production, which would suggest some environmental efficiency gains in lowering emissions in agriculture (Figure 10.3). To some extent the environmental efficiency gains in reducing ammonia emissions associated with changes in livestock numbers and manure management, is also likely to be attributed to changes in inorganic fertiliser use and management.

The environmental efficiency gains in reducing the level and rate of release of agricultural ammonia emissions over the past decade, can be primarily linked to the uptake of improved technologies and farm management practices, as well as incentives to lower emissions provided by a range of policies introduced by OECD countries. Increasing numbers of farmers are adopting technologies (e.g. covered manure storage facilities) and practices that are helping to reduce emissions, such as precision fertiliser application. The adoption of these technologies and practices has partly been associated with the use of various policy instruments, for example, regulations on the storage and spreading of manure, and payments for manure storage (OECD, 2012). Emission rates and the effectiveness of mitigation practices, however, are affected by conditions outside the control of farmers, including temperature, rainfall and wind.

In terms of progress towards achieving the emission targets set for 2010 under the *Gothenburg Protocol*, there is a varied picture across OECD countries (Table 10.2). By 2008-10 many countries had reduced their emissions to meet their target levels under the Protocol. But some countries will need to achieve further emission reductions to attain the 2010 target, especially **Denmark** and **Finland,** while another group of countries are much closer to reaching the target (**Germany, Spain** and **Switzerland**). However, all these countries are encouraging widespread adoption of farm nutrient management practices and implementing programmes that seek to reduce ammonia emissions (OECD, 2008).

References

Eilers, W., R. MacKay, L. Graham and A. Lefebvre (eds.) (2010), "Environmental Sustainability of Canadian Agriculture", *Agri-Environmental Indicator Report Series*, Report No. 3, Agriculture and Agri-Food Canada, Ottawa, Canada, *http://publications.gc.ca/collections/collection_2011/agr/A22-201-2010-eng.pdf.*

EMEP, *website of the* Co-Operative Programme for Monitoring and Evaluation of the Long-Range Transmission of Air Pollutants in Europe (EMEP), *www.emep.int/index_data.html.*

European Environment Agency (2012), *Emissions of Acidifying Substances (CSI 001) – Assessment Published Dec. 2011,* website content: see *www.eea.europa.eu/data-and-maps/indicators/emissions-of-acidifying-substances-version-2/assessment-1.*

European Environment Agency (2000), "Environmental Signals 2000", *Environmental Assessment Report,* No 6, Copenhagen, Denmark.

National Atmospheric Deposition Program, United States Ammonia Monitoring Network, website *http://nadp.sws.uiuc.edu/amon.*

OECD (2012), *Water Quality and Agriculture: Meeting the Policy Challenge*, OECD Publishing, *www.oecd.org/agriculture/water.*

OECD (2008), *Environmental Performance of Agriculture in OECD Countries Since 1990*, OECD Publishing, *www.oecd.org/tad/sustainable-agriculture/agri-environmentalindicators.htm.*

UNECE (2012), *Long-Term Financing of the Co-operative Programme for Monitoring and Evaluation of the Long-Range Transmission of Air Pollutants in Europe* (EMEP), website page: *www.unece.org/env/lrtap/emep_h1.html.*

UNECE/EMEP (2009), *Joint EMEP/CORINAIR Atmospheric Emission Inventory Guidebook*, European Environment Agency, Copenhagen, Denmark, *www.eea.europa.eu/publications/emep-eea-emission-inventory-guidebook-2009*.

Chapter 11

Greenhouse gas emissions: Climate change

This chapter reviews the environmental performances of agriculture in OECD countries related to greenhouse gas emissions. It provides a description of the policy context (issues and main challenges), definitions for the agri-environmental indicators presented, and elements related to concepts, interpretations, links to other indicators, as well as measurability and data quality. The chapter then describes the main trends of the agri-environmental indicators, using available data covering the period 1990-2010 and based on a set of tables and figures.

This chapter examines agricultural greenhouse gas emissions (methane CH_4, nitrous oxide N_2O and carbon dioxide CO_2), which together with the previous chapter on ammonia emissions (Chapter 10), examines how agricultural activities impact on air quality. The environmental impacts of these agricultural emissions should be viewed in the broader context of other pollution sources (e.g. industry, transport) and considered in terms of the chemical reactions between different air pollutants in the atmosphere ("multi-pollutants", e.g. sulphur dioxide, carbon dioxide) and the resultant effects on the environment ("multi-effects", e.g. acidification, eutrophication) (Figure 10.1).

11.1. Policy context

The issue

The relations between agriculture and climate change are complex compared to many other economic activities because agriculture: contributes to emissions of greenhouse gas (GHGs); provides a carbon sink function under certain management practices; while agriculture is also subject to the impacts of climate change. The issues of agricultural carbon sinks and the impact of climate change on agriculture are outside the scope of this report.

Most OECD countries are committed to GHG emission targets (from 1990 levels) to be achieved by the 2008-12 timeframe, but there are no specific reduction targets set for methane or nitrous oxide, and agriculture, like other sectors, does not have specific commitments under the United Nations Framework Convention on Climate Change (UNFCCC). However, all OECD countries are developing agricultural climate change programmes that aim to reduce GHGs, promote carbon sinks, and make agriculture more resilient to climate change impacts.

While these programmes vary they mainly involve a mix of approaches, such as the use of regulations supported by investment subsidies and farm advice to encourage adoption of certain practices that will lower GHG emission rates and promote carbon sinks. Other policies are acting indirectly to address climate change in agriculture, such as programmes providing incentives for less intensive use of agricultural land, improved nutrient management, as well as removing land from agricultural production (afforestation, land conservation programmes, extensive use of grassland), which can also contribute to carbon sequestration.

Main challenges

The main challenge in relation to agriculture and agricultural GHG emissions is in reducing the overall level and rate of emission release per unit volume of agricultural production. While there are other challenges in agriculture in relation to climate change, such as encouraging carbon sequestration and improving agriculture's resilience to climate change, these are not measured through the GHG indicator in this report as they are outside the scope of this study. The challenge of lowering GHG emissions should also take into account the potential synergies or trade-off with the other environmental issues in the agricultural sector, such as water pollution and biodiversity conservation.

11.2. Indicators

Definitions

The indicator related to agricultural greenhouse gas emissions measures changes in:

- Gross total agricultural greenhouse gas emissions (methane and nitrous oxide but excluding carbon dioxide).

Concepts, interpretation, limitations and links to other indicators

Agriculture's link to greenhouse gas (GHG) emissions and climate change is complex. While the sector is a contributor of GHGs to the atmosphere, some components of agricultural production systems, (i.e. soils) can act as carbon sinks depending on how they are managed (OECD, 2008). Certain agricultural biomass feedstocks can provide a neutral carbon source of renewable energy. Moreover, while farming is a source of greenhouse gases, principally methane (CH_4) and nitrous oxide (N_2O), which are part of the primary driving force behind climate change, equally climate change may also impact on farm production. Impacts and adaptation to climate change may cause shifts in crop types and cropping patterns in many OECD countries, but these issues are outside the scope of this report.

While overall the UNFCCC Inventories provide a robust and internationally comparable dataset, there are a number of limitations. National emission estimates made by individual member countries may vary depending on which factors are included in their own calculations. Agricultural sources of CO_2 emissions are in many countries aggregated with emissions from forestry and fisheries, and hence, not included in the figures and tables of this chapter. In addition, assumptions made in agricultural GHG emission calculations simplify complex agricultural systems introducing uncertainty into the estimate of GHG emissions. Also the country emission data shown in this chapter may vary between countries depending on which specific UNFCCC methodology a country has used in its calculations (further details of UNFCCC methodologies are available at: *http://unfccc.int/ghg_data/ghg_data_unfccc/items/4146.php*).

Agricultural GHG emissions are linked to indicators of nitrogen balances (Chapter 4), ammonia (Chapter 10), energy use (Chapter 6), and soil carbon stocks, as driving forces in terms of their consequences (or state) for global warming and impacts on climate change. Agriculture's response to reducing GHGs has been partly through increasing the production and use of renewable energy, improving energy efficiency and also, by lowering emissions through improved nutrient and soil management practices and improving the efficiency of livestock production.

Measurability and data quality

Major agricultural sources of CH_4 and N_2O, such as enteric fermentation (a process during livestock digestion, where microbes in the digestive system ferment food consumed by the animal), livestock manure, fertiliser and saturated agricultural soils (e.g. wet paddy fields), are covered by the agricultural module of the UNFCCC Inventories. The Inventory excludes data for CO_2 (mainly fossil fuel combustion in agriculture), which is included in the energy module.

UNFCCC inventories are the main source of data on GHG emissions used in this chapter. These provide a dataset in accordance with the methodology of the Intergovernmental Panel on Climate Change (IPCC) Guidelines for National Greenhouse Gas Inventories. The UNFCCC data are comparable as they cover most OECD countries,

Figure 11.1. **Agricultural gross greenhouse gas emissions, OECD countries, 1990-2010**

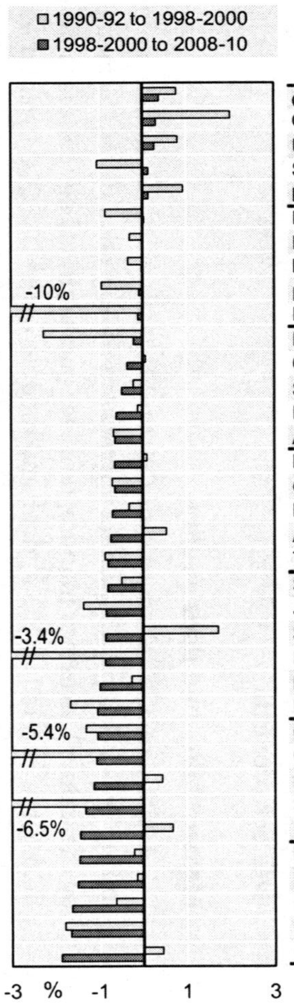

Average annual percentage change: thousand tonnes CO_2 equivalent

□ 1990-92 to 1998-2000
■ 1998-2000 to 2008-10

-10% // ; -3.4% // ; -5.4% // ; -6.5% //

-3 % -1 1 3

	Average			Change in agricultural GHG emissions[1]				Share of agriculture in national total GHG emissions	Share in total OECD agriculture GHG emissions
	Thousand tonnes, CO_2 equivalent			Thousand tonnes, CO_2 equivalent	Average annual % change	Thousand tonnes, CO_2 equivalent	Average annual % change	%	%
	1990-92	1998-2000	2008-10	1990-92 to 1998-2000		1998-2000 to 2008-10		2008-10	2008-10
Chile[2]	12 285	13 070	13 425	785	0.8	354	0.4	9	1
Canada	46 885	54 943	56 672	8 058	2.0	1 729	0.3	8	5
United States	392 450	418 180	429 546	25 731	0.8	11 366	0.3	6	36
Switzerland	6 091	5 603	5 675	-488	-1.0	72	0.1	11	0
New Zealand	30 777	33 128	33 556	2 351	0.9	429	0.1	46	3
Finland	6 222	5 809	5 823	-413	-0.9	14	0.0	8	0
Iceland	677	661	660	-16	-0.3	-1	0.0	14	0
Mexico[3]	46 739	45 486	45 424	-1 253	-0.3	-62	0.0	7	4
Israel[4,5]	2 681	2 581	2 565	-100	-0.9	-16	-0.1	3	0
Estonia	3 210	1 377	1 362	-1 833	-10.0	-15	-0.1	7	0
Poland	43 412	36 133	35 345	-7 279	-2.3	-788	-0.2	9	3
OECD	**1 220 878**	**1 226 552**	**1 182 632**	**5 674**	**0.1**	**-43 920**	**-0.4**	**8**	**100**
Slovenia	2 111	2 073	1 974	-38	-0.2	-99	-0.5	10	0
France	103 022	102 038	96 029	-985	-0.1	-6 009	-0.6	18	8
Austria	8 530	8 081	7 577	-449	-0.7	-504	-0.6	9	1
Norway	4 544	4 575	4 288	31	0.1	-287	-0.6	8	0
Germany	77 798	73 524	68 868	-4 274	-0.7	-4 656	-0.7	7	6
Luxembourg	750	730	681	-19	-0.3	-50	-0.7	6	0
Australia	86 151	89 870	83 370	3 720	0.5	-6 500	-0.7	15	7
Turkey	30 153	28 115	25 955	-2 037	-0.9	-2 160	-0.8	7	2
Sweden	8 895	8 545	7 882	-349	-0.5	-663	-0.8	12	1
Japan	31 202	27 943	25 665	-3 259	-1.4	-2 278	-0.8	2	2
Spain	37 301	42 733	39 185	5 432	1.7	-3 549	-0.9	10	3
Hungary	12 225	9 234	8 458	-2 991	-3.4	-776	-0.9	12	1
EU15	**426 951**	**417 972**	**378 073**	**-8 978**	**-0.3**	**-39 900**	**-1.0**	**10**	**32**
Denmark	12 321	10 765	9 689	-1 556	-1.7	-1 076	-1.0	15	1
Greece	11 298	10 166	9 148	-1 132	-1.3	-1 018	-1.0	7	1
Czech Republic	13 960	8 938	8 026	-5 022	-5.4	-913	-1.1	6	1
Portugal	8 138	8 430	7 521	292	0.4	-909	-1.1	10	1
Slovak Republic	6 029	3 518	3 079	-2 511	-6.5	-439	-1.3	7	0
Ireland	19 752	20 835	17 999	1 083	0.7	-2 836	-1.5	28	2
Belgium	11 763	11 555	9 967	-208	-0.2	-1 588	-1.5	8	1
Italy	41 086	40 574	34 844	-512	-0.2	-5 730	-1.5	7	3
United Kingdom	57 271	54 449	46 167	-2 822	-0.6	-8 282	-1.6	8	4
Netherlands	22 804	19 739	16 692	-3 065	-1.8	-3 046	-1.7	8	1
Korea[5]	22 349	23 152	19 514	803	0.4	-3 638	-1.9	..	2

Figure 11.1. **Agricultural gross greenhouse gas emissions, OECD countries, 1990-2010** *(cont.)*

Notes: Countries are ranked in descending order according to average annual percentage change 1998-2000 to 2008-10.
The statistical data for Israel are supplied by and under the responsibility of the relevant Israeli authorities. The use of such data by the OECD is without prejudice to the status of the Golan Heights, East Jerusalem and Israeli settlements in the West Bank under the terms of international law.

1. Gross GHG emissions from agriculture include emissions of CH_4 and N_2O but exclude CO_2 emissions.
2. Data for 2008-10 refer to 2005-07.
3. Data for 2008-10 refer to 2004-06.
4. Data for 1990-92 refer to the year 1996; and data for 1998-2000 refer to the year 2000.
5. Data for 2008-10 refer to 2007-09.

Source: UNFCCC (2012), website of the *UNFCCC Greenhouse Gas Inventory Database, http://ghg.unfccc.int*; and national data (for Chile, Israel, Korea and Mexico).

StatLink http://dx.doi.org/10.1787/888932793243

Figure 11.2. **Agricultural greenhouse gas emissions and agricultural production volume, OECD countries, 1990-2010**

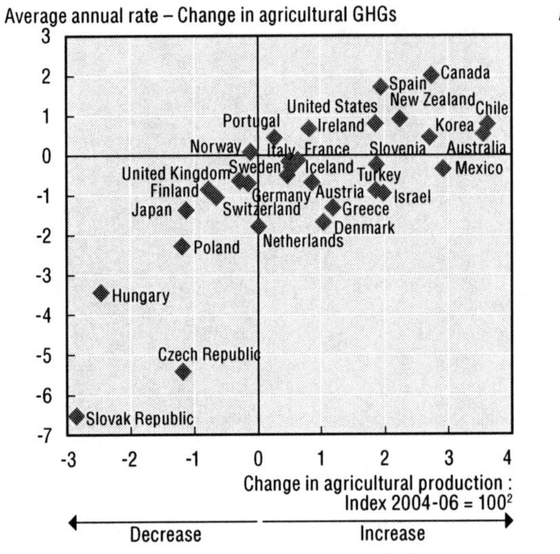

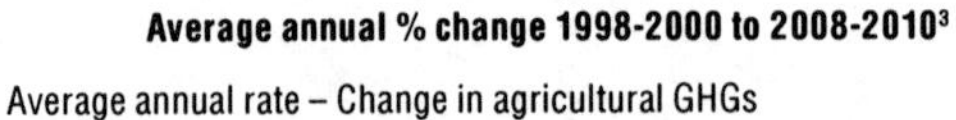

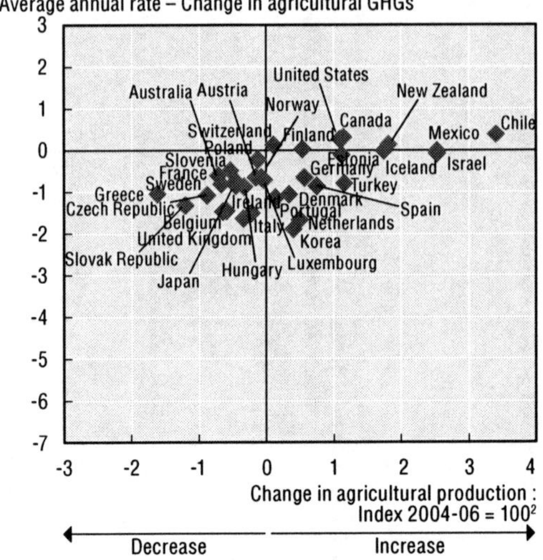

Notes: The statistical data for Israel are supplied by and under the responsibility of the relevant Israeli authorities. The use of such data by the OECD is without prejudice to the status of the Golan Heights, East Jerusalem and Israeli settlements in the West Bank under the terms of international law.

1. For the period 1990-92 to 1998-2000, Estonia (-5% per annum agricultural production volume and -10% per annum agricultural greenhouse gas emissions), is not represented. Agricultural production index is not available for Belgium and Luxembourg. Israel agricultural greenhouse gas emissions data for 1990-92 refer to the year 1996; and for 1998-2000 refer to the year 2000.
2. The FAO indices of agricultural production show the relative level of the aggregate volume of agricultural production for each year in comparison with the base period 2004-06. They are based on the sum of price weighted quantities of different agricultural commodities produced after deductions of quantities used as seed and feed weighted in a similar manner. The resulting aggregate represents, therefore, disposable production for any use except as seed and feed. All the indices at the country, regional and world levels are calculated by the Laspeyres formula. Production quantities of each commodity are weighted by 2004-06 average international commodity prices and summed for each year. To obtain the index, the aggregate for a given year is divided by the average aggregate for the base period 2004-06. Due to technical reasons it is not possible to provide an OECD or EU average.
3. For the period 1998-2000 to 2008-10, agricultural greenhouse gas emissions: Chile: data for 2008-10 refer to 2005-07; Mexico: data for 2008-10 refer to 2004-06; Israel and Korea: data for 2008-10 refer to 2007-09.

Source: FAOSTAT (2012); UNFCCC (2012), website of the *UNFCCC Greenhouse Gas Inventory Database, http://ghg.unfccc.int*; and national data (for Chile, Israel, Korea and Mexico).

StatLink http://dx.doi.org/10.1787/888932793262

while for **Chile, Israel, Korea** and **Mexico** national data are used. Emissions of CH_4 and N_2O are converted to carbon dioxide (CO_2) equivalents using weights (Global Warming Potentials). However, CO_2 emissions from the upstream and downstream agro-food sectors, such as fertiliser and pesticide manufacturing, energy use, transportation and processing are not included in this report, because the focus of the GHG indicator here is on primary agriculture.

A number of OECD countries are beginning to monitor carbon sequestration in agricultural soils and report these to the UNFCCC. The UNFCCC inventories will in the future categorise carbon sequestration in agricultural soils separately from soil emissions in accordance with the new LULUCF (i.e. land use, land-use change and forestry) reporting requirements. The UNFCCC also collects data on emissions from land use changes, but these data are not included here as it is not possible to extract data explicit to farm land use change (i.e. farm land converted to/from other uses).

Table 11.1. **Gross greenhouse gas emissions, OECD countries, 1990-2010**

Carbon dioxide equivalent

Type of GHG	Gross OECD total emissions[1] (million tonnes)			Share of each gas in OECD total[1] (%)			GHG Emissions from agriculture[2] (million tonnes)			Share of each gas in total emissions from agriculture (%)			Share of agriculture in OECD total of each gas (%)		
	1990-92	1998-2000	2008-10	1990-92	1998-2000	2008-10	1990-92	1998-2000	2008-10	1990-92	1998-2000	2008-10	1990-92	1998-2000	2008-10
Carbon dioxide (CO_2)	11 657	12 645	12 515	81	82	83	..	..	..	..	..	..	..	..	..
Methane (CH_4)	1 603	1 582	1 544	11	10	10	605	605	594	50	49	50	38	38	38
Nitrous oxide (N_2O)	918	896	783	6	6	5	615	621	588	50	51	50	67	69	75
Others: (HFCs, PFCs, SF6)	234	272	282	2	2	2	..	..	..	..	..	..	..	..	..
Total	**14 412**	**15 395**	**15 124**	**100**	**100**	**100**	**1 220**	**1 226**	**1 182**	**100**	**100**	**100**	**8**	**8**	**8**

1. Gross OECD total emissions, excluding land use, land-use change and forestry (LULUCF). Gross OECD Total emissions do not include Korea for CO_2, CH_4 and N_2O; and do not include Israel and Korea for others (HFCs, PFCs, SF6).
2. Gross GHG emissions from agriculture include emissions of CH_4 and N_2O but exclude CO_2 emissions.

Source: UNFCCC (2012), website of the *UNFCCC Greenhouse Gas Inventory Database*, *http://ghg.unfccc.int*; and national source for Chile, Israel, Korea and Mexico.

StatLink http://dx.doi.org/10.1787/888932793528

11.3. Main trends

Over the decade of the 2000s, total gross OECD agricultural GHG emissions decreased by 0.4% annually compared to a small increase of 0.1% per annum over the 1990s, leading to an overall reduction of nearly 44 million tonnes of CO_2 equivalent over the decade (Figure 11.1). During the period 1998-2000 to 2008-10, very few countries registered an increase in GHG emissions, notably **Canada, Chile, New Zealand** and the **United States**, but at a significantly lower growth rate than in the 1990s. The **United States** GHG emissions increased by over 11 million tonnes of CO_2 equivalent over the last decade (Figure 11.1).

The **EU15** experienced an overall diminution of -1% per annum in agricultural GHG emissions between 1998-2000 and 2008-10, allowing a saving of nearly 40 million tonnes of CO_2 equivalents, with **France**, **Germany**, **Italy**, and the **United Kingdom** contributing a major part of the EU reduction. Other EU countries also experienced large reductions in emissions, for example the **Netherlands** and **Spain**, while agricultural production increased (Figures 11.1 and 11.2). In some countries such as **Australia**, **Ireland**, **Korea**, **Portugal** and **Spain**, the upward trend in agricultural GHG emissions over the 1990s was reversed to a declining trend over the last decade.

For the EU transition countries (**Czech and Slovak Republics**, **Estonia**, **Hungary**, **Poland** and **Slovenia**), GHG emissions have declined continuously since the 1990s (Figure 11.1). This can be mainly explained by the sharp contraction of the agricultural sectors of these countries in the transition to a market economy during the 1990s, followed by a period of agricultural recovery over the 2000s, when the rate of GHG emission release, although continuing to decline, was at a slower rate than over the 1990s, except for the **Slovak Republic**.

Overall the share of agriculture in total OECD GHG emissions was 8% in 2008-10, but averaged much higher for nitrous oxide (N_2O) and methane (CH_4), at 75% and 38% respectively (Figure 11.1 and Table 11.1). The relative contribution of agriculture in the total of national GHG emissions varies across countries, with five countries having a share 15% or higher in 2008-10 (**Australia, Denmark, France, Ireland** and **New Zealand**), although the contribution of these countries to the total OECD agricultural GHG emissions was low except for **Australia** and **France**. Together the **EU15** and the **United States** accounted for 68% of OECD agricultural GHG emissions in 2008-10.

The contribution of agriculture to global GHG emissions mainly came from two gases: nitrous oxide (N_2O) and methane (CH_4), which accounted for around 50% each of OECD total agricultural emissions in CO_2 equivalents in 2008-10, with these shares changing little since the early 1990s (Table 11.1). Nitrous oxide emissions largely derive from the application of fertilisers, manure waste, crop residues and the cultivation of organic soils (which also result in carbon dioxide emissions), while methane emissions mainly result from livestock enteric fermentation and manure. In **Japan** and **Korea**, rice paddy production is also an important source of methane emissions.

Trends in agricultural GHG emissions are principally determined by changes in agricultural production, in particular, livestock production leading to changes in methane (CH_4) emissions and crop production linked to fertiliser use affecting changes in nitrous oxide (N_2O) emissions. Relating trends in total agricultural production to trends in agricultural GHG emissions over the period 1990-2010, indicates that overall there has been an improvement in environmental efficiency of agricultural GHG emissions (i.e. trends in GHG emissions changing at a lower rate than the corresponding change in agricultural production) (Figure 11.2).

It would appear that the environmental efficiency improvements in GHG emissions over the 2000s have been marked for **Canada**, **Chile**, **Estonia**, **Iceland**, **Israel**, **Mexico**, **New Zealand** and the **United States** (growth or near stable GHG emissions trends compared to faster annual growth rates in agricultural production); **Denmark**, **Korea**, **Netherlands**, **Spain** and **Turkey** (reductions in GHG emissions but increases in agricultural production); and **Belgium**, **Ireland**, **Italy** and the **United Kingdom** (the reduction in GHG emissions has been greater than the decrease in agricultural production) (Figure 11.2).

The environmental efficiency gains in reducing the level and rate of release of agricultural GHG emissions over the past decade, as with ammonia, can be primarily linked to the uptake of improved technologies and farm management practices, as well as incentives to lower emissions provided by a range of policies introduced by OECD countries. Increasing numbers of farmers are adopting technologies (e.g. improving feed efficiency and livestock growth rates to reduce methane emissions) and practices that are helping to reduce emissions, such as precision fertiliser application (lowering nitrous oxide emissions).

The adoption of these technologies and practices has partly been associated with the use of various policy instruments, for example, regulations on livestock housing to limit GHG emissions, and payments for biodigesters to capture and produce methane as a source of renewable energy and indirectly replace energy sources such as coal. In **Canada**, for example, between 1981 and 2006, adoption of improved management practices by the beef and pork industries have led to a nearly 40% reduction in GHG emissions per unit of liveweight produced and for the dairy industry 20% per kilogramme of milk produced (Eilers et al., 2010).

References

Eilers, W., R. MacKay, L. Graham and A. Lefebvre (eds.) (2010), "Environmental Sustainability of Canadian Agriculture", *Agri-Environmental Indicator Report Series*, Report No. 3, Agriculture and Agri-Food Canada, Ottawa, Canada, *http://publications.gc.ca/collections/collection_2011/agr/A22-201-2010-eng.pdf.*

OECD (2008), *Environmental Performance of Agriculture in OECD Countries Since 1990*, OECD Publishing, *www.oecd.org/tad/sustainable-agriculture/agri-environmentalindicators.htm.*

Chapter 12

Methyl bromide: Ozone depletion

This chapter reviews the environmental performances of agriculture in OECD countries related to methyl bromide. It provides a description of the policy context (issues and main challenges), definitions for the agri-environmental indicators presented, and elements related to concepts, interpretations, links to other indicators, as well as measurability and data quality. The chapter then describes the main trends of the agri-environmental indicators, using available data covering the period 1990-2010 and based on a set of tables and figures.

12.1. Policy context

The issue

Methyl bromide is used as a fumigant in the agriculture, horticulture and food sectors, which can be harmful to human health and soil biodiversity because of its high toxicity, but is destructive as an ozone-depleting substance, which is of concern for human health and the environment. The Parties to the *Montreal Protocol on Substances that Deplete the Ozone Layer* agreed in 1997 to a global phase-out schedule for methyl bromide. Under the schedule, developed countries had to reduce methyl bromide use by 25% by 1999, 50% by 2001, 70% by 2003 and 100% by 2005, compared to 1991 levels (OECD, 2008; UNEP).

Developing countries (i.e. Article 5 member countries under the *Montreal Protocol*) started a freeze on use in 2002 at average 1995-98 levels, and need to achieve a 20% reduction by 2005 and 100% by 2015 (UNEP). Among OECD countries, **Chile**, **Korea**, **Mexico** and **Turkey** are included under Article 5 of the *Montreal Protocol*.

Main challenges

The challenge for agriculture is to reduce methyl bromide use to meet the agreed reduction levels under the *Montreal Protocol*, leading to its eventual elimination from use in agriculture (Table 12.1).

12.2. Indicators

Definitions

The indicator related to agricultural methyl bromide use measures the change in:

- Methyl bromide use, expressed in tonnes of ozone depleting substance equivalents.

Concepts, interpretation, limitations and links to other indicators

Methyl bromide is a fumigant that has been used for more than 50 years in the agriculture, horticulture and food sectors. It is used to control soil insects, diseases, nematodes and mites in open fields and greenhouses and for pests associated with the storage of food commodities, such as grains. This fumigant has also been used for plant quarantine and pre-shipment protection (OECD, 2008).

While methyl bromide has the advantage of being a low cost fumigant that affects a broad spectrum of pests, it is harmful to human health and soil biodiversity because of its high toxicity. Methyl bromide is also an ozone-depleting substance that is more destructive to the ozone layer than many other ozone depleting substances. Ozone depletion hinders the activities of stratosphere ozone layers. The ozone layer prevents harmful ultraviolet (UV-B) rays from reaching the earth which can cause damage to crop production, forest growth, and human and animal health (OECD, 2008).

Table 12.1. **Methyl bromide use, world and OECD countries, 1991-2010**

	Tonnes of ozone depletion potential (ODP tonnes) from methyl bromide use				% change (1991 to 2010)	Montreal protocol agreed % reduction level in 2005 from 1991 base year
	1991[1]	2008	2009	2010		
OECD[2]	33 453	2 622	2 108	2 103	-94	x
United States	15 317	1 817	1 363	1 633	-89	100
EU15	11 530	165	0	0	-100	100
Japan	3 664	236	167	149	-96	100
Israel	2 148	360	541	286	-87	100
Australia	422	25	20	21	-95	100
Canada	120	20	17	14	-88	100
Poland	120	0	0	0	-100	100
New Zealand	81	0	0	0	-100	100
Hungary	32	0	0	0	-100	100
Czech Republic	6	0	0	0	-100	100
Norway	6	0	0	0	-100	100
Slovak Republic	6	0	0	0	-100	100
Article 5 countries[3]	Average 1995-1998[1]	2008	2009	2010	% change 1995-98 to 2010	Montreal protocol agreed % reduction level in 2005 (2015) from 1995-98 base year
Mexico	1 131	820	745	668	-41	20 (100)
Turkey	480	0	0	0	-100	20 (100)
Chile	213	164	165	162	-24	20 (100)
Korea	0	0	0	0	-100	20 (100)
	1991	2008	2009	2010	% change 1991-2010	
World use of ODP products (tonnes)	894 253	43 452	48 906	43 895	-95	
Share of OECD methyl bromide use in world total use of ODP products (%)	4	6	4	5		
World total methyl bromide use (ODP tonnes)	38 665	5 984	4 914	4 185	-89	
Share of OECD in world total methyl bromide use (%)	87	44	43	50		

Notes: Countries are ranked from highest to lowest user of methyl bromide in 1991.
The statistical data for Israel are supplied by and under the responsibility of the relevant Israeli authorities. The use of such data by the OECD is without prejudice to the status of the Golan Heights, East Jerusalem and Israeli settlements in the West Bank under the terms of international law.
The following countries have either already eliminated use of methyl bromide by 1991 or have not used the substance (excluding EU15 countries): Estonia, Iceland, Korea, Slovenia and Switzerland.
1. 1991 base period for non-Article 5 countries under the *Montreal Protocol* and 1995-98 for Article 5 countries.
2. OECD excludes Chile, Korea, Mexico and Turkey.
3. Article 5 countries under the *Montreal Protocol.*

Source: UNEP Ozone Secretariat (2012), *http://ozone.unep.org/new_site/en/ozone_data_tools_access.php.*

StatLink *http://dx.doi.org/10.1787/888932793547*

As an environmental driving force, the methyl bromide use indicator links to the state (and changes in) of the ozone layer. OECD countries are obliged to *respond* in eliminating methyl bromide use based on the schedule agreed under the *Montreal Protocol.*

Measurability and data quality

Methyl bromide use data are collected by the Parties to the *Montreal Protocol*, and reported to the Ozone Secretariat, which is hosted by the United Nations Environment Programme (UNEP). Parties report production, import and export quantities in metric tonnes and the Secretariat calculates the weighted consumption using each substance's ozone-depleting potential (ODP). The ODP is a relative index indicating the extent to which a chemical product may cause ozone depletion (OECD, 2008). The ODP coefficient of methyl bromide is 0.6, and ODP tonnes are calculated as follows:

Methyl bromide ozone depleting potential (ODP tonnes) =
Methyl bromide use (tonnes) × Ozone Depletion Potential Coefficient

The data of methyl bromide use for the **EU15** member states were reported to the UNEP as aggregated data for the EU15 in accordance with the *Montreal Protocol*. It should be noted that methyl bromide use for the purpose of quarantine and pre-shipment is exempt from the phase-out programme and the use data for these purposes are not reported to the UNEP, and hence, excluded from the OECD database. Thus, for those countries reporting zero use of methyl bromide by primary agriculture in this chapter, they might be using the pesticide in the agro-food sector, for quarantine and pre-shipment use.

12.3. Main trends

Most OECD countries have achieved the reduction level targets for methyl bromide specified under the *Montreal Protocol* up to 2010 (Table 12.1). Some OECD countries, however, up to 2010 were still using methyl bromide for critical uses beyond the agreed phase-out date of 2005 under the *Montreal Protocol*, notably the **United States**, and to a lesser extent **Israel** and **Japan** (Table 12.1). This group of countries have all made a significant reduction in methyl bromide use by around 90% or more by 2010 compared to 1991 levels (1991 is the base period under the *Montreal Protocol*) (Table 12.1). Since 2012, there has been a complete ban on methyl bromide use in **Israel**.

World use of total ODP products declined by 95% during the period 1991 to 2010, with the reduction in methyl bromide slightly less at 89%, and little change in its share in world total ODP use at around 5% over this period (Table 12.1). Moreover, OECD countries' share of world total methyl bromide use declined from over 87% in 1991 to 50% in 2010, which stemmed from a reduction in OECD methyl bromide use of over 90% (excluding Article 5 countries) over this period (Table 12.1).

Soil fumigation treatment, especially for horticultural crops, accounts for about three-quarters of global methyl bromide use (OECD, 2008). In addition, methyl bromide is used for the storage of durable commodities (e.g. grains and timber) and perishable commodities (e.g. fresh fruit and vegetables, cut-flowers), and the disinfestations of structures (e.g. buildings, ships and aircraft). (OECD, 2008).

Reductions in methyl bromide use have been achieved by a combination of government regulations and changes in the market, as well as pressure from non-governmental organisations and the activities of private companies. Moreover, some countries have adopted a more stringent phase-out schedule than required under the *Montreal Protocol*, including efforts to develop alternatives (OECD, 2008).

For the four OECD countries – **Chile**, **Korea**, **Mexico** and **Turkey** – covered under Article 5 of the *Montreal Protocol*, the use of methyl bromide also decreased between 1995-98 to 2010 (Table 12.1). **Korea** and **Turkey** have already eliminated methyl bromide use completely, while **Chile** and **Mexico** have made reductions in use beyond that required under the Protocol, although they both still have to make substantial cuts in use to meet the total elimination of methyl bromide by 2015.

For a few OECD countries, the phase-out schedule for methyl bromide has posed a technical challenge in terms of finding alternatives. In view of these technical difficulties, the *Montreal Protocol* allows the Parties to apply for **Critical Use Exemptions** (CUEs) when there are no feasible alternatives, in addition to the existing exemption for use in quarantine and pre-shipment purposes.

The CUEs are intended to give farmers, fumigators and other users of methyl bromide additional time to develop substitutes (OECD, 2008). In 2009 and 2010, the Parties to the *Montreal Protocol* agreed to CUEs for 2012, including for **Australia**, **Canada**, **Israel**, **Japan** and the **United States** (UNEP). The CUE process has allowed those sectors facing technical challenges in finding alternatives to continue agricultural production while conducting research and implementing alternative treatments. However, granting CUEs may impede the effectiveness of the phase out schedule under the *Montreal Protocol* and act as a disincentive for CUE countries to seek alternatives (OECD, 2008).

References

OECD (2008), *Environmental Performance of Agriculture in OECD Countries Since 1990*, OECD Publishing, *www.oecd.org/tad/sustainable-agriculture/agri-environmentalindicators.htm*.

UNEP (United Nations Environment Programme), Ozone Secretariat website: *http://ozone.unep.org*.

Chapter 13

Biodiversity: Farmland bird populations and agricultural land cover

This chapter reviews the environmental performances of agriculture in OECD countries related to biodiversity. It provides a description of the policy context (issues and main challenges), definitions for the agri-environmental indicators presented, and elements related to concepts, interpretations, links to other indicators, as well as measurability and data quality. The chapter then describes the main trends of the agri-environmental indicators, using available data covering the period 1990-2010 and based on a set of tables and figures.

13.1. Policy context

The issue

The Convention on Biological Diversity (CBD, 2002) defines agricultural biodiversity at levels from genes to ecosystems that are involved or impacted by agricultural production (Box 13.1). Agricultural biodiversity is distinct in that it is largely created, maintained, and managed by humans through a range of farming systems from subsistence to those using a range of biotechnologies and extensively modified terrestrial ecosystems. In this regard, agricultural biodiversity stands in contrast to "wild" biodiversity which is most valued *in situ* and as a product of natural evolution (OECD, 2008).

Box 13.1. **Defining agricultural biodiversity**

Drawing on the CBD definition of biodiversity, agricultural biodiversity can be defined in terms of (OECD, 2001):

1. ***Genetic diversity:*** the number of genes within domesticated plants and livestock species and their wild relatives.
2. ***Species diversity:*** the number and population of wild species (flora and fauna) both dependent on, or impacted by, agricultural activities, including soil biodiversity and effects of non-native species on agriculture and biodiversity.
3. ***Ecosystem diversity:*** populations of domesticated and wild species and their non-living environment (e.g. climate), which make up an *agro-ecosystem* and is in contact with other ecosystems (i.e. forest, aquatic, steppe, rocky and urban). The agro-ecosystem consists of a variety of habitats limited to an area where the ecological components are quite homogenous and are cultivated, such as extensive pasture or an orchard, or are uncultivated but within a farming system, such as a wetland.

OECD countries employ a variety of policies and approaches to reconcile the need to enhance farm production, drawing on plant and livestock genetic resources, and yet reduce harmful biodiversity impacts, especially on wild species (e.g. birds) and ecosystems (e.g. wetlands). In addition most OECD countries are signatories to international agreements of significance for agro-biodiversity conservation, such as the *Convention on Biological Diversity*; the *Convention on the Conservation of Migratory Species of Wild Animals*; and the *Ramsar Convention* for the protection of wetlands. To better understand the complexity of agri-biodiversity linkages and with the aim of developing a set of indicators that can capture this complexity, the OECD has developed an Agri-Biodiversity Indicators Framework (ABF) (Figure 13.1).

The ABF recognises three key aspects in agri-biodiversity linkages, which operate at varying spatial scales from the field to the global level. First, an agro-ecosystem provides both food and non-food commodities, and environmental services (e.g. scientific,

Figure 13.1. **OECD Agri-Biodiversity Indicators Framework**

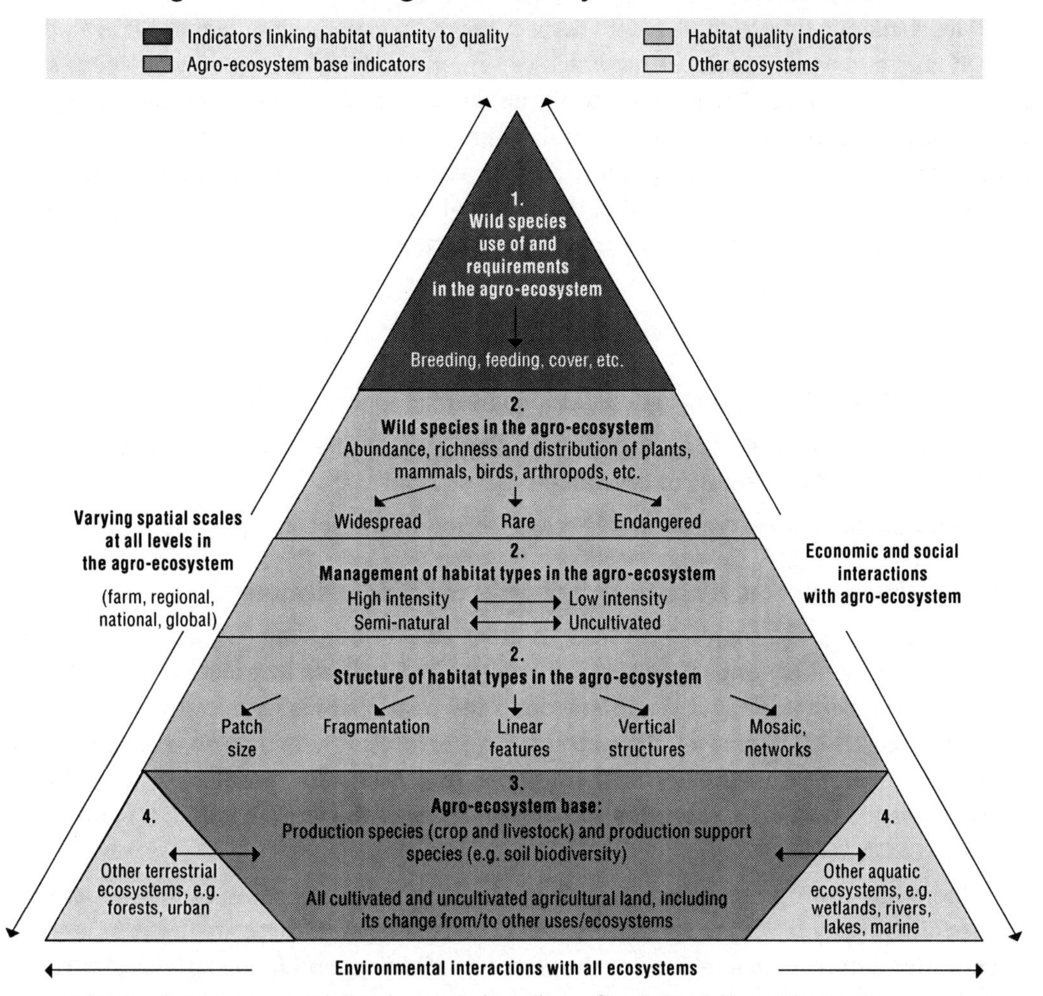

Source: OECD (2003), *Agriculture and Biodiversity: Developing Indicators for Policy Analysis*, OECD Publishing, *www.oecd.org/tad/sustainable-agriculture/agri-environmentalindicators.htm*.

recreational, ecological). Second, the agro-ecosystem consists of plant and animal communities (domesticated crops and livestock, and wild species), which interact with the economic and social aspirations of farming. Third, the agro-ecosystem is linked to other ecosystems, both terrestrial (e.g. forests) and aquatic (e.g. wetlands), especially in terms of the effects of farming practices on other ecosystems but also the effects of these ecosystems on agriculture.

Within an agro-ecosystem the ABF highlights a hierarchical structure of three layers, from which OECD has developed some indicators discussed in this and other chapters of the report, as well as in previous OECD work (OECD, 2008). The first layer is the production base of agriculture, in particular, its use of *genetic resources* (plants and livestock). Indicators of agricultural genetic resources are not included in this report, but were previously examined in the first edition of *At a Glance* (OECD, 2008). A second layer consists of the structure (e.g. field mosaic, linear features) and management (i.e. variety of farming practices and systems) of *habitats* within the agro-ecosystem, which impacts on the third and final layer. This layer covers the abundance, richness and distribution of *wild species* either dependent on or impacted by agricultural activities.

Main challenges

The challenge for agriculture with respect to biodiversity is significant as it is a major user of land and water resources on which certain genetic resources and wild species are highly dependent (Table 2.1). Efforts toward the conservation of birds on farmland may also help contribute to broader biodiversity goals of protecting the diversity of wild species and ecosystems associated with agriculture. The main challenge for agriculture, more specifically with regard to farmland birds, is to maintain or restore breeding populations, while developing a profitable business to expand agricultural production.

13.2. Indicators

Definitions

The indicators concerning agricultural biodiversity include change in:

- Populations of a selected group of breeding bird species that are dependent on agricultural land for nesting or breeding.
- Agricultural land cover types – arable crops, permanent crops and pasture areas.

Concepts, interpretation, limitations and links to other indicators

Agriculture is the major land user in most OECD countries (Figure 3.6). As such, agriculture has a direct impact on species' habitats and indirect impacts on the existence of the species themselves, but the interactions and relationships that control impacts are complex (Figure 13.1). Moreover, the consequences of farming activities on wild species are especially important in those OECD countries (e.g. **Australia**, **Mexico**) which have a "megadiversity" status (i.e. countries with a high share of the world's wild flora and fauna species) (OECD, 2008).

There has been progress made with methods to calculate some indicators of wild species biodiversity related to agriculture (OECD, 2003). However, there are few comparative, quantitative data available relating to the status of wild flora and fauna species associated with agriculture across OECD countries. The notable exceptions are bird populations, although more data are also becoming available on butterflies, but for flora the data are much poorer (OECD, 2008).

Birds can act as "indicator species" providing a barometer of the health of the environment. Being close to or at the top of the food chain, they reflect changes in ecosystems rather rapidly compared to other species. The farmland bird index is an average trend in a group of species suited to track trends in the condition of farmland habitats. In general, a decrease in the index means that the balance of bird species population trends are negative, representing biodiversity loss. If it is constant, there is no overall change. An increase in the index implies that the balance of bird species trends are positive, implying that biodiversity loss has halted.

An increasing farmland bird index may, or may not, always equate to an improving situation in the environment. It could in extreme cases be the result of expansion of some bird species at the cost of others. In all cases, detailed analysis must be conducted to interpret accurately the indicator trends, while the composite index trend of farmland birds can hide important changes for individual species.

While bird populations are impacted by agricultural activities, such as the loss of habitats on farmland, many other factors external to farming also affect population dynamics, for example, changes in populations of "natural" predators, the weather, and over longer periods climate change. A further issue is defining primary agricultural habitat, as some bird species may use farmland as a feeding area but breed in an adjoining forest, while changes in adjacent ecosystems may themselves affect bird species using farmland. Moreover, some farmland bird species are migratory and impacts on the population can arise from changes in their migratory ranges (summer or winter) and also changes along their migratory route (OECD, 2003; and 2008).

The farmland bird index indicator is also based solely on breeding farmland bird populations and assumes that they effectively represent the biodiversity of farmland. But this may lead to the exclusion of some species, such as wintering birds, for example, swans and geese which migrate to overwinter in Ireland every year (as well as many other parts of Europe) from the arctic and boreal nesting areas. Most of these species rely on managed grassland for feeding throughout the winter period.

Tracking changes in the area of agricultural semi-natural habitats, can provide information on the extent of land that is subject to relatively "low intensity" farming practices, such as wooded pastures and extensive grasslands with little, if any, fertilisers and pesticides used in their management, and relatively undisturbed by machinery operations (especially during the nesting season) or not farmed at all, such as fallow land (uncultivated habitats on farmland, such as hedges). A major difficulty in assessing changes in semi-natural habitats on agricultural land is their definition in terms of what constitutes "semi-natural" across different farming systems and countries, although international agreements, such as the CBD, are beginning to address this issue.

A further limitation is that at present, for most countries, data of semi-natural habitats are collected at fairly broad levels of aggregation which impairs analysis of potential impacts on biodiversity. Indeed, this chapter uses the very broad category of permanent pasture as a proxy for semi-natural habitats, subject to all the caveats mentioned here.

Overall changes in agricultural land use (Chapter 3), and land cover discussed in this chapter, as well as farming management practices and systems are key drivers on farmland bird populations. In addition, the intensity of input use, especially nutrients (Chapter 4), pesticides (Chapter 5) and water resources (Chapter 8), as well as water quality (Chapter 9), are also important driving forces that link to the state and trends of farmland bird populations and responses in terms of conservation programmes as part of broader agri-biodiversity management plans.

Measurability and data quality

The farmland bird indicator used in this chapter mainly draws on the Birdlife International's (BI) Pan European Common Bird Monitoring Scheme (European Bird Census Council), as well as national bird monitoring programmes. An important difference in these various national indices are the number and type of species included, ranging from 8 to 36 bird species, to reflect varying national situations, and the various methods used to derive the indices (see the detailed notes in the OECD website). BI treats data with statistical techniques that enable calculation of national species' indices and their combination into supranational indices for species, weighted by estimates of national population sizes. Weighting allows for the fact that some countries hold different proportions of each species' population. Supranational indices for species are then combined (on a geometric scale) to create multi-species indicators, fixed (for the purpose of presentation) to a value of 100 in 2000. Farmland bird population indices are currently only available for **OECD European countries**, **Canada** and the **United States** (only grassland species), but efforts are being made under the Biodiversity Indicators Partnership (BIP) to develop a global wild bird index building on national data (see *www.bipindicators.net/wbi*).

Data on the main types of agricultural land use (i.e. arable and permanent crops and permanent pasture) are most commonly collected by census, more regularly for arable crops, but frequently only every 10 years for the area of permanent pasture. Despite increasing scientific knowledge of the ecological functions of biodiversity, habitat monitoring and assessment systems are, for most OECD countries, poor in terms of disaggregated time series. Many countries, however, are beginning to make an effort to monitor changes in semi-natural and uncultivated habitat areas on farmland as part of a broader national biodiversity management plan. Even so, by examining changes in areas of permanent pasture on farmland, as well as linking this to the overall change in total agricultural land area (Chapter 3), plus using other scattered information on land use and cover changes, some tentative conclusions can be drawn as to likely impacts for bird populations and other flora and fauna using farmland as a habitat.

13.3. Main trends

Farmland birds

Trends in OECD ***farmland bird populations*** declined continuously over the period from 1990 to 2010 for almost all countries. The decrease in farmland bird populations, however, was less pronounced over the 2000s compared to more rapid reductions in the 1990s in most cases (Figures 13.2 and 13.3). Nearly all OECD countries witnessed a reduction of the index of farmland birds, declining by over 15% from 1998-2000 to 2008-10 for **Austria**, **Belgium**, **Denmark**, **Germany**, the **Netherlands** and **Norway.** However, there were small increases in the index for **Estonia**, **Ireland** and the **United States** (grassland species only). For a few countries the rate of reduction in farmland bird populations was more rapid over the last decade compared to the 1990s, including for **Denmark**, **Germany** and the **Netherlands** (Figures 13.2 and 13.3).

In describing the broad trends for OECD countries, this needs to be treated with caution as all OECD countries outside of Europe, except **Canada** and the **United States**, are missing from the dataset. Partial evidence from the non-European OECD countries suggest that trends in farmland bird populations are following a similar pattern to those of European countries, with an overall decline over the past two decades, with perhaps the rate of decline slowing over the last decade. The **Canadian** wildlife habitat capacity on farmland indicator showed that between 1986 to 2006 the average habitat capacity on farmland declined due to loss of natural and semi-natural land cover and the intensification of agricultural operations (Eilers et al., 2010). In the **United States**, the index of grassland bird species has been in long term decline since the mid-1960s, but since around the mid-2000s, there has been a marked improvement (Figures 13.2 and 13.3).

Australian studies point to declines in bird species on grassland habitats and with pressure on bird populations from land clearing by agriculture, particularly livestock grazing, although this has eased considerably in recent years (State of the Environment 2011 Committee, 2011). In a few countries, notably **Japan** and **Korea,** where paddy rice agriculture is widely practised, this system of farming can provide an important habitat for birds. This depends, however, on the management practices used in paddy systems, and at

Figure 13.2. **Farmland bird index, OECD countries, 1990-2010**

% annual average change

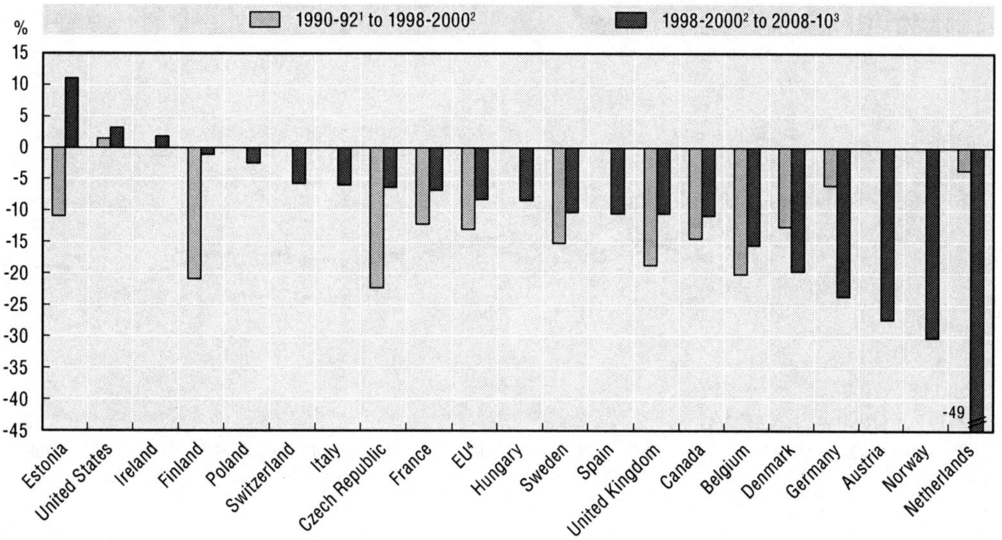

Notes: Aggregated index of population trend estimates of a selected group of breeding bird species that are dependent on agricultural land for nesting or feeding. For Canada and the United States these are only grassland breeding birds.

Countries are ranked in descending order according to average annual percentage change 1998-2000 to 2008-10.

1. Data for the 1990-92 average equal 1991-93 average for Germany.
2. Data for the 1998-2000 average equal 1999-2001 average for Hungary and Switzerland; and 2000-02 average for Italy and Poland.
3. Data for the 2008-10 average equal 2004-06 average for Estonia; 2005-07 average for the United States, 2007-09 average for Hungary; and 2006-08 average for Belgium, Denmark, Finland, France, Germany, Ireland, Spain, Sweden, Switzerland and United Kingdom.
4. The EU aggregate figure is an estimate based on the following 17 member states: Austria, Belgium, the Czech Republic, Denmark, Estonia, Finland, France, Germany, Hungary, Ireland, Italy, Latvia, the Netherlands, Poland, Spain, Sweden, and United Kingdom.

Source: Statistical Office of the European Union (EUROSTAT, see *http://epp.eurostat.ec.europa.eu*) and various national sources (Austria, Germany, Hungary, Ireland, Italy, Netherlands, Norway, Poland and Switzerland); Environment Canada (2012), *The State of Canada's Birds*; and United States (2009), *North American Bird Conservation Initiative. State of the Birds 2009*, Department of Interior, Washington, DC.

StatLink http://dx.doi.org/10.1787/888932793281

present Korea and Japan have no national farmland bird monitoring system, although recent research has revealed that more than 30% of native avian species in these two countries use rice fields as habitats (Fujioka et al., 2010).

In other OECD countries that do not monitor trends in farmland bird populations, notably **Chile**, **Israel**, **Mexico**, **New Zealand** and **Turkey**, with the expansion and intensification of agricultural production (Figure 3.7), it is possible that farmland bird populations have been subject to increasing pressure and likely declines in populations in common with the recent trends in other countries.

The slowdown in the rate of decline of farmland bird populations over the 2000s compared to the 1990s (and earlier decades), has been partly associated with: efforts beginning in the early 1990s to introduce agri-environmental schemes in many countries aimed at encouraging semi-natural land conservation on farms (e.g. field margins, buffer strips near rivers and wetlands); changes in farm management practices, such as increasing the area under conservation tillage which has increased feed supplies for birds and other wild species; reductions in nutrient surpluses (Figures 4.1 and 4.2) and pesticide sales (Figure 5.1) for most countries, lowering toxic effects on birds and their food supply (e.g. worms, insects); and changes in land use (discussed below).

Figure 13.3. **Trends in the farmland birds index, OECD countries, 1990-2010**

Index 2000 = 100

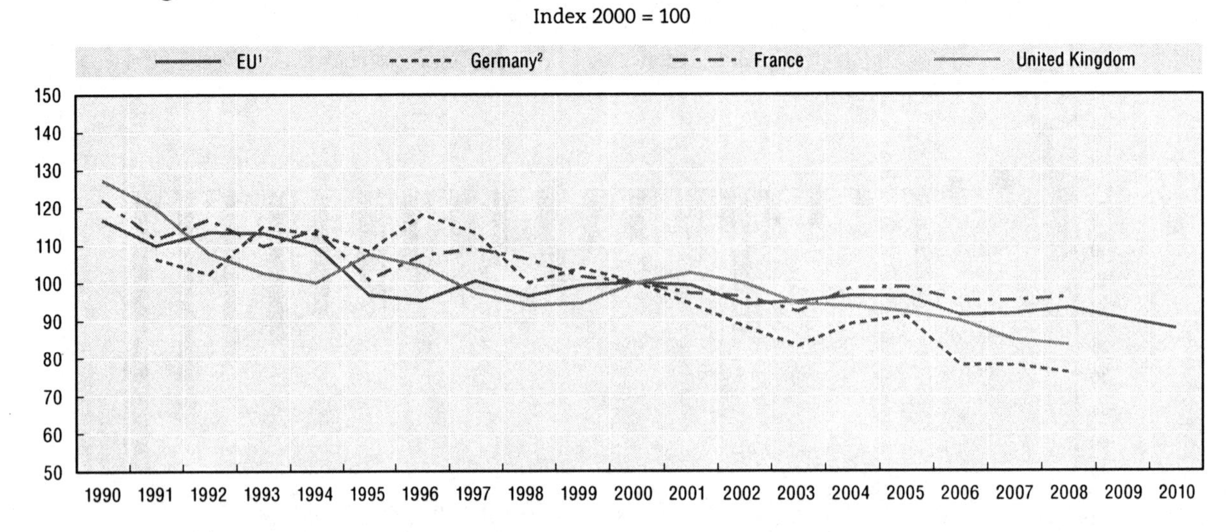

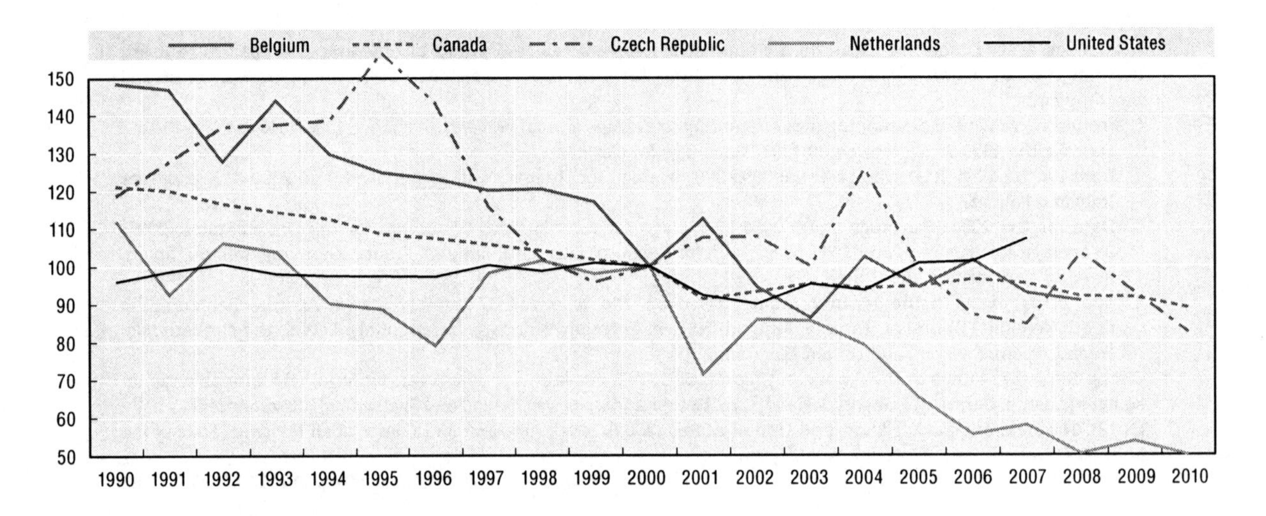

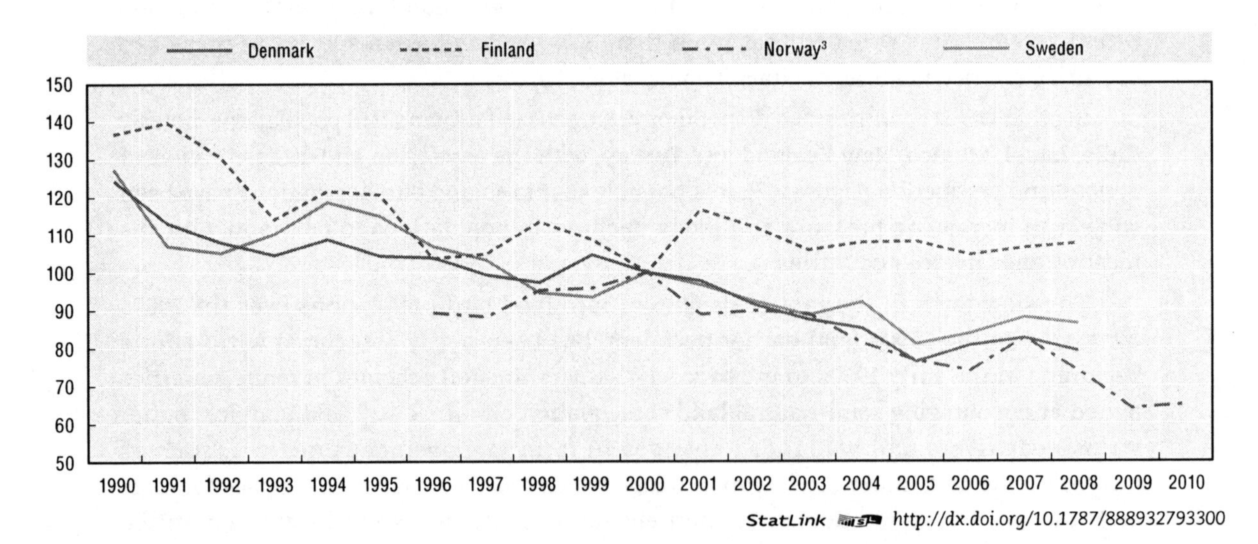

StatLink http://dx.doi.org/10.1787/888932793300

Figure 13.3. **Trends in the farmland birds index, OECD countries, 1990-2010** *(cont.)*

Index 2000 = 100

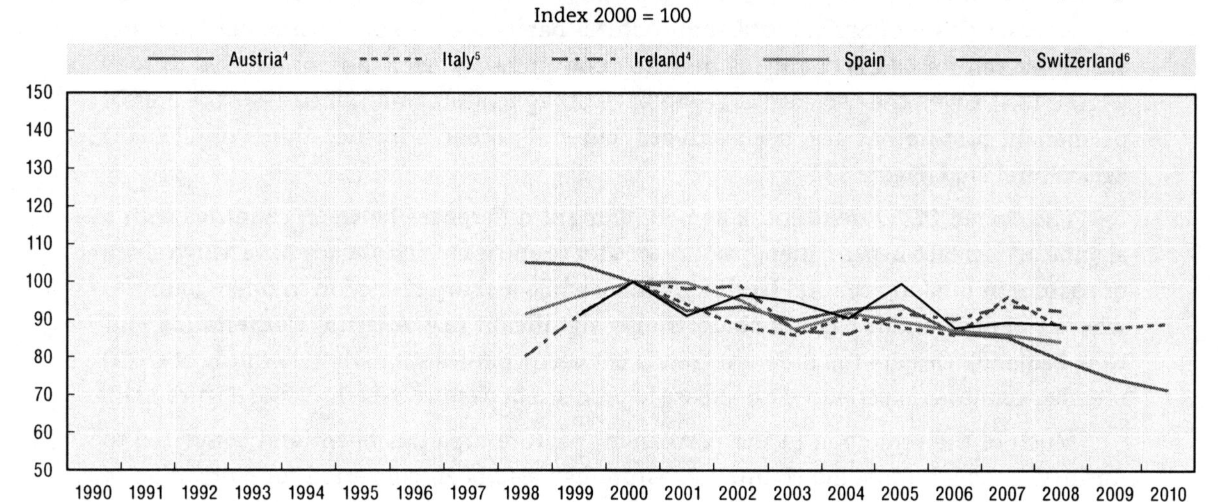

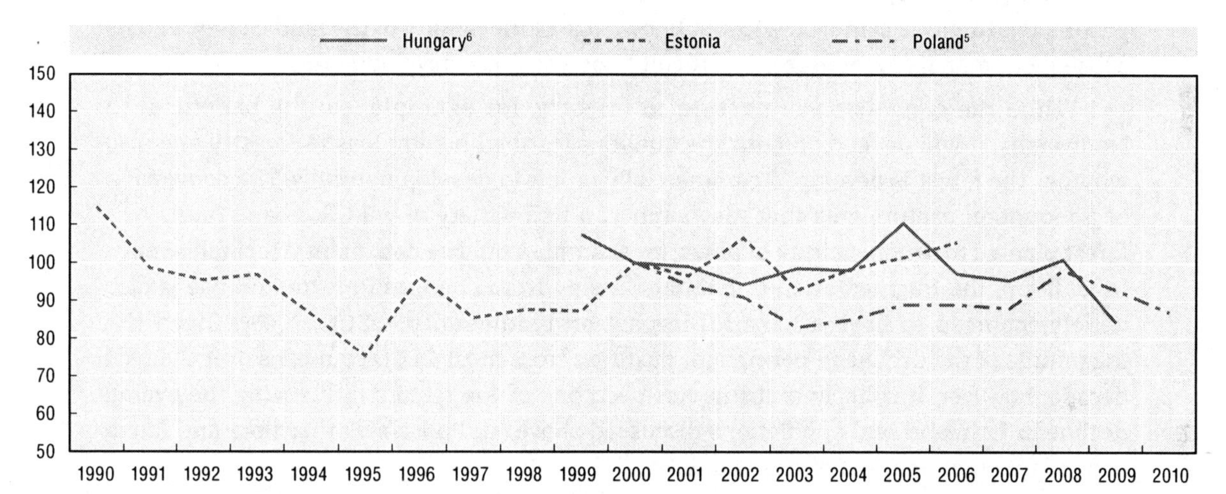

Notes: Aggregated index of population trend estimates of a selected group of breeding bird species that are dependent on agricultural land for nesting or breeding. For Canada and United States, these are only grassland breeding birds.

1. The EU aggregate figure is an estimate based on the following 17 member states: Austria, Belgium, the Czech Republic, Denmark, Estonia, Finland, France, Germany, Hungary, Ireland, Italy, Latvia, the Netherlands, Poland, Spain, Sweden, and United Kingdom.
2. Values are not available for Germany for 1990.
3. For 1990-95, values are not available for Norway.
4. For 1990-97, values are not available for Austria and Ireland.
5. For 1990-99, values are not available for Italy and Poland.
6. For 1990-98, values are not available for Hungary and Switzerland.

Source: Statistical Office of the European Union (EUROSTAT, see *http://epp.eurostat.ec.europa.eu*) and various national sources (Austria, Germany, Hungary, Ireland, Italy, Netherlands, Norway, Poland and Switzerland); Environment Canada (2012), *The State of Canada's Birds*; and United States (2009), *North American Bird Conservation Initiative. State of the Birds 2009*, Department of Interior: Washington, DC.

StatLink http://dx.doi.org/10.1787/888932793300

Despite these positive improvements toward bird conservation on farmland across many OECD countries, the further intensification of agriculture and removal of natural and semi-natural habitats in some regions of the OECD, continues to exert pressure on bird populations and other flora and fauna associated with farming. It is also noticeable that bird species dependent on other habitats, notably forestry, have not experienced the same rate of decline as in farmland bird species (OECD, 2011a).

Agricultural land cover

A major share of agricultural semi-natural habitats consists of ***permanent pasture***, which overall for OECD countries declined continuously over the period 1990 to 2010 (Figure 13.4). Given the decrease in the total OECD agricultural land since 1990, the area of permanent pasture has also been reduced, but still accounts for two-thirds of all OECD agricultural land (Figure 13.5).

The overall OECD trend mask some important differences between countries with a significant increase in permanent pasture area for countries which already have a high share of pasture in total agricultural land (e.g. **Chile**), while a sharp reduction in other countries where the permanent pasture share is also significant (e.g. **Austria**, **Netherlands** and **New Zealand**). Despite the different trends between countries, nearly two thirds of OECD member countries experienced a decrease in permanent pasture over the 2000s (Figure 13.4).

Much of the reduction in the permanent pasture area has been land converted to forestry, although for some countries pasture has also been converted for cultivation of arable and permanent crops (e.g. **Finland** and the **Netherlands**) (OECD, 2008). Interpreting the consequences for farmland birds and other wildlife species of changes in permanent pasture land areas is complex. Without knowledge of the quality of the land change and its subsequent management makes it difficult to assess these developments.

While the conversion of pasture to forestry, for example, can be beneficial to biodiversity, it will depend on both the quality of farmed habitat loss to forestry and also whether the forest is developed commercially or left to develop naturally. The conversion of a mountain pasture area that may support a rich variety of wild flora and fauna to a forest planted to a monoculture of pines, for example, could be detrimental for biodiversity. In addition, the fragmentation of habitats arising from changes in farmland use is also widely reported to have a harmful impact on biodiversity (OECD, 2008). Given the magnitude of the decline in permanent pasture across most OECD countries over the past decade, however, it is likely that this has been one of the factors influencing the overall decline in farmland bird populations discussed above, and possibly other flora and fauna dependent on permanent pasture land.

More generally, the assessment of land use changes both between agriculture and other uses of land (e.g. forestry, urban use) and within agriculture (e.g. between pasture and arable crops), is incomplete in this report. This is because of the paucity of datasets to provide a complete analysis of these changes, including data on how different land types are managed and thereby influence the wild flora and fauna that use farmland as a habitat. Recent OECD research (OECD, 2011b) over the period 2000-06 reveals that for some countries conversion of land for urban land use has mainly originated from agricultural land (e.g. **Czech** and **Slovak Republics**, **Germany**) and for other countries (e.g. **Finland**, **Norway** and **Slovenia**) primarily from forests. At the same time land covered originally by forest and natural vegetation was converted to mainly agricultural use in, for example **Finland**, **Portugal**, **Spain** and **Turkey**.

Until these data gaps are adequately filled it will be difficult to undertake an evaluation of policies that address biodiversity conservation in agriculture or alert policy makers and the wider public to detrimental trends or success stories of agricultural biodiversity conservation. Further scientific research is also required to better understand the environmental consequences of different land conversions (e.g. agriculture to forestry) and the quality of those changes (e.g. how the land converted from one use to another is

Figure 13.4. **Permanent pasture and arable and permanent crop land area, OECD countries, 1990-2010**

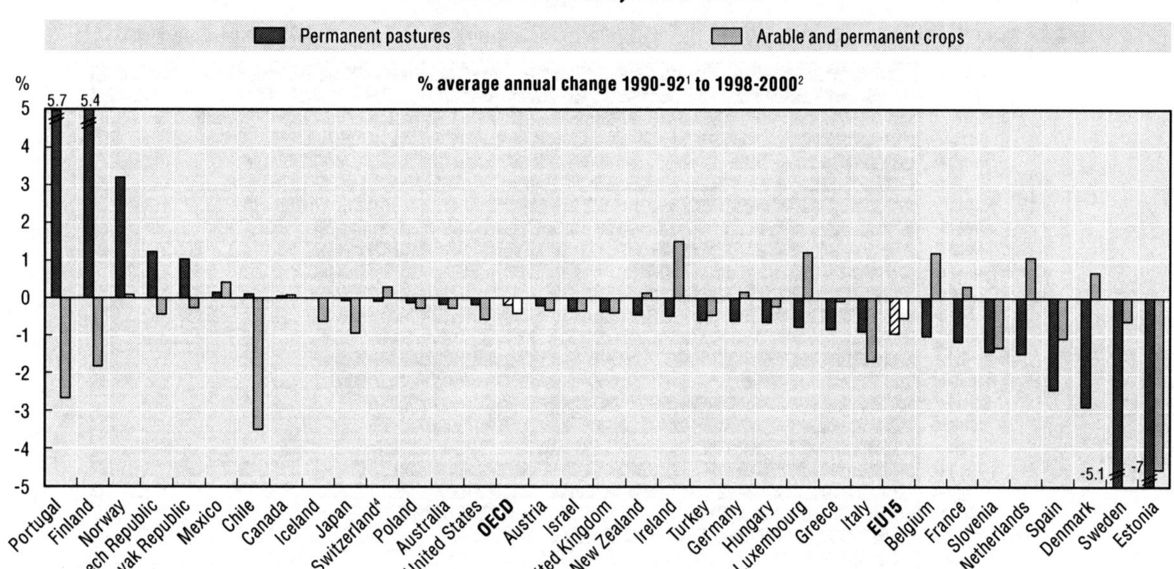

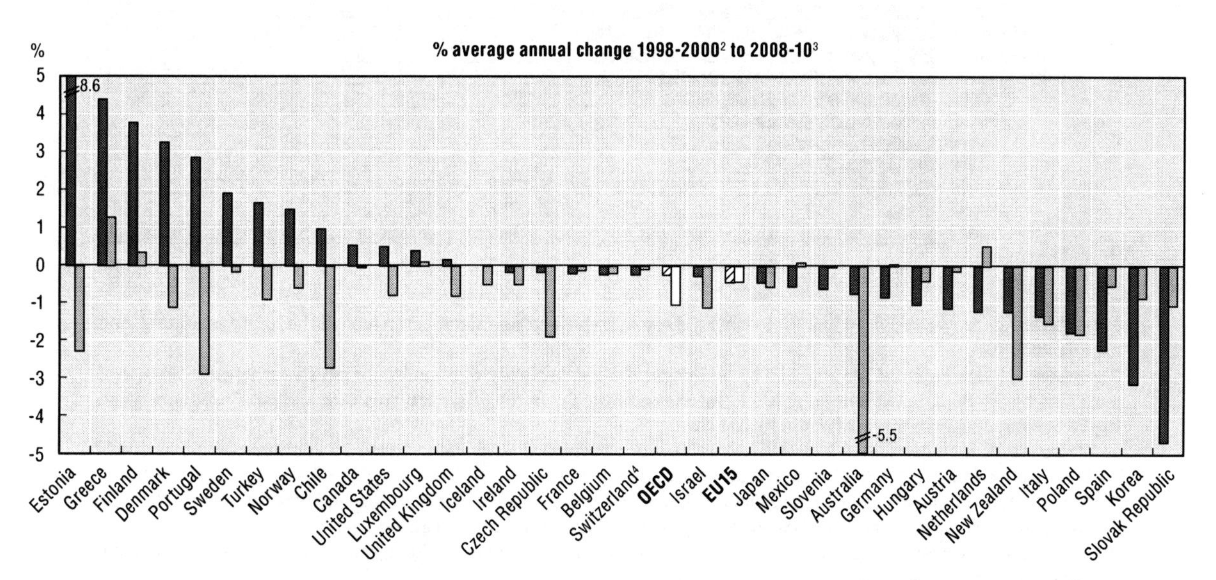

Notes: Countries are ranked in terms of highest to lowest average annual % change for 1990-92 to 1998-2000 and for 1998-2000 to 2008-10. The statistical data for Israel are supplied by and under the responsibility of the relevant Israeli authorities. The use of such data by the OECD is without prejudice to the status of the Golan Heights, East Jerusalem and Israeli settlements in the West Bank under the terms of international law.

1. Data for 1990-92 average equal to the 1991-92 average for Slovenia: the year 1990 for Greece; and the year 1991 for Switzerland.
2. Data for 1998-2000 average equal to the 1999-2001 average for Austria; and the year 2000 for Greece.
3. Data for 2008-10 average equal to the 2007-09 average for Austria, Canada, Denmark, Iceland, Israel, Korea and Mexico; the 2006-08 average for Chile; the 2007-08 average for Italy; and the year 2007 for Greece.
4. In the case of Switzerland, data refer to Utilised Agricultural Area (hectares), including arable and permanent cropland, but excluding summer pastures.

Source: FAOSTAT (2012), *http://faostat.fao.org*; Statistical Office of the European Union (EUROSTAT, see *http://epp.eurostat.ec.europa.eu*); and national data for Australia, Canada, Iceland, Japan, Korea, Turkey and United States.

StatLink *http://dx.doi.org/10.1787/888932793319*

managed), not only for wild species and ecosystems, but also for soil erosion, water quality, flood control, and carbon sequestration, for example.

Figure 13.5. **Arable cropland, permanent cropland, permanent pasture and other agricultural land share in the total agricultural land area, OECD countries, 2008-10**

Share of arable and permanent crop area | Share of permanent pasture area | Share of other land[1]

Country	Share of arable and permanent crop area	Share of permanent pasture area
Finland	98	1
Korea[2]	97	2
Denmark[2]	92	8
Japan	87	13
Sweden	85	15
Norway	83	17
Hungary	82	18
Greece[2]	80	20
Poland	78	20
Czech Republic	74	26
Israel[2]	73	27
Spain	72	28
Germany	72	28
Italy[2]	71	27
Slovak Republic	71	27
Canada[2]	67	33
Estonia	66	33
France	66	34
EU	64	36
Belgium	63	37
Turkey	63	37
Netherlands	56	43
Portugal	51	48
Luxembourg	48	52
Austria[2]	45	55
Slovenia	42	58
Switzerland[3]	42	58
United States	40	60
United Kingdom	35	65
OECD	33	67
Mexico[2]	27	73
Ireland	20	80
Chile[2]	11	89
Australia	6	94
Iceland[2]	5	95
New Zealand	3	97

0 20 40 60 80 100 %

Notes: Countries are ranked from highest to lowest share of arable and permanent crop area in the total agricultural land area for 2008-10.

The statistical data for Israel are supplied by and under the responsibility of the relevant Israeli authorities. The use of such data by the OECD is without prejudice to the status of the Golan Heights, East Jerusalem and Israeli settlements in the West Bank under the terms of international law.

1. "Other land" includes fallow and other categories of agricultural land not included by countries as either arable cropland, permanent cropland or land under permanent pasture.
2. Data for 2008-10 average equal to the 2007-09 average for Austria, Canada, Denmark, Iceland, Israel, Korea and Mexico; the 2006-08 average for Chile; the 2007-08 average for Italy; and year 2007 for Greece.
3. In the case of Switzerland, data refer to Utilised Agricultural Area (hectares), including arable and permanent cropland, but excluding summer pastures.

Source: FAOSTAT (2012), *http://faostat.fao.org*; Statistical Office of the European Union (EUROSTAT, see *http://epp.eurostat.ec.europa.eu*); and national data for Australia, Canada, Iceland, Japan, Korea, Turkey, and United States.

StatLink http://dx.doi.org/10.1787/888932793338

References

Biodiversity Indicators Partnership website, *www.bipindicators.net/wbi*.

Convention on Biological Diversity (2002), *Agricultural Biodiversity – Introduction and Background*, Secretariat of the Convention on Biological Diversity, United Nations Environment Programme, *www.biodiv.org/programmes/areas/agro*.

Eilers, W., R. MacKay, L. Graham and A. Lefebvre (eds.) (2010), "Environmental Sustainability of Canadian Agriculture", *Agri-Environmental Indicator Report Series*, Report No. 3, Agriculture and Agri-Food Canada, Ottawa, Canada, *http://publications.gc.ca/collections/collection_2011/agr/A22-201-2010-eng.pdf*.

Environment Canada (2012), *The State of Canada's Birds*, *www.stateofcanadasbirds.org*.

European Bird Census Council (EBCC), *www.ebcc.info*.

FAOSTAT, Land use database, *http://faostat.fao.org*.

Fujioka, M., S.D. Lee, M. Kurechi and H. Yoshida (2010), "Bird Use of Rice Fields in Korea and Japan", *Waterbirds*, Vol. 33, Special Publication 1, pp. 8-29.

North American Bird Conservation Initiative (NABCI), *www.nabci-us.org/stateofthebirds.htm*.

OECD (2011a), *Towards Green Growth: Monitoring Progress OECD Indicators*, OECD Publishing, *www.oecd.org/greengrowth*.

OECD (2011b), *OECD Regions at a Glance 2011*, OECD Publishing.

OECD (2008), *Environmental Performance of Agriculture in OECD Countries Since 1990*, OECD Publishing, Paris, France, *www.oecd.org/tad/sustainable-agriculture/agri-environmentalindicators.htm*.

OECD (2003), *Agriculture and Biodiversity: Developing Indicators for Policy Analysis*, OECD Publishing, *www.oecd.org/tad/sustainable-agriculture/agri-environmentalindicators.htm*.

OECD (2001), *Environmental Indicators for Agriculture: Vol. 3*, OECD Publishing.

State of the Environment 2011 Committee (2011), *Australia State of the Environment 2011*, Department of Sustainability, Environment, Water, Population and Communities, Australian Government, Canberra, Australia.

ANNEX A

Use of indicators for policy monitoring and evaluation

Policy relevance is a key criterion in the selection and development of agri-environmental indicators (AEIs). This section provides illustrative examples of how the OECD's AEIs are being used in policy analysis and monitoring across a range of countries, institutions and researchers.

Denmark: Policies to reduce nutrient pollution

Since 1985, **Denmark** has implemented **a set of national measures to reduce agricultural nutrient pollution** of water systems, especially to avoid eutrophication of coastal water (Box A1). These measures are in conformity with the EU's *Nitrate and Water Framework Directives*, and in part, funded through the Common Agricultural Policy, with Box A1 showing use of the nutrient balance indicator.

Box A1. **Policy measures to reduce agricultural nutrient pollution of the environment in Denmark**

Danish policy actions	Policy measures imposed
1985: Nitrogen (N) and phosphorus (P) action plan to reduce N and P pollution	• Minimum 6 months slurry storage capacity. • Ban on slurry spreading between harvest and 15 October on soil destined for spring cropping. • Maximum stock density equivalent to 2 livestock unit (LU) ha^{-1} (1 LU corresponds to one large dairy cow). • Various measures to reduce runoff from silage clamps and manure heaps. • A floating barrier (natural crust or artificial cover) mandatory on slurry tanks.
1987: The first action plan for the Aquatic Environment (AP-I), aiming to halve N-losses and reduce P-losses by 80%	• Minimum 9 months slurry storage capacity. • Ban on slurry spreading from harvest to 1 November on soil destined for spring crops. • Mandatory fertiliser and crop rotation plans. • Minimum proportion of area to be planted with winter crops. • Mandatory incorporation of manure within 12 hours of spreading.
1991: Action plan for a Sustainable Agriculture	• Ban on slurry spreading from harvest until 1 February, except on grass and winter rape. • Obligatory fertiliser budgets. • Maximum limits on the plant available N applied to different crops, equal to the economic optimum. • Statutory norms for the proportion of manure N assumed to be plant available (Pig slurry: 60%, cattle slurry: 55%, deep litter: 25%, other types: 50%).
1998: The second action plan for the Aquatic Environment (AP-II)	• Subsidies to establish wetlands. • A reduction of the stock density maximum to 1.7 LU ha^{-1}. • The statutory norms for the proportion of manure N assumed to be plant available were increased from 1999 (pig slurry: 65%, cattle slurry: 60%, deep litter: 35%, other types: 55%). • Maximum limits on the application of plant available N to crops reduced to 10% below the economic optimum. • Mandatory 6% of the area with cereals, legumes and oil crops to be planted with catch crops.

Box A1. **Policy measures to reduce agricultural nutrient pollution of the environment in Denmark** *(cont.)*

Danish policy actions	Policy measures imposed
2000: AP-II mid-term evaluation and enforcement	• Further tightening of the statutory norms for the proportion of assumed plant-available N in manure. From 2001; pig slurry: 70%, cattle slurry: 65%, deep litter: 40%, other types: 60%. From 2002 pig slurry: 75%, cattle slurry: 70%, deep litter: 45%, other types: 65%.
2004: The third action plan for the Aquatic Environment (AP-III) AP-III is closely related to the EU Water Framework Directive and the EU Habitat Directive. N-leaching must be reduced by further 13% by 2015	• Further tightening of the request for catch crops. • Establishment of buffer zones along streams and around lakes to reduce discharge of P. • A tax of DKK 4 kg^{-1} (EUR 0.54 kg^{-1}) mineral P in feed. • Evaluations of the effect of AP-III will be carried out in 2008 and 2011. • Based on the evaluations further initiatives will be implemented if necessary.
2008: Evaluation of AP-III	
2009: Political agreement on initiating AP-IV; also called Green growth General reduction targets for the aquatic environment are estimated, and regional objectives are set for individual water bodies	• Further tightening of the request for catch crops; in total 140 000 hectares with catch crops is needed. • Ban on soil tillage in the autumn before spring crops. • Ban on re-establishment of fodder grass in the autumn. • Work initiated to evaluate the possibilities of introducing tradable leaching quotas.

Implementation of the Danish Action Plans, together with the support and regulatory framework of the EU's Nitrate and Water Framework Directives and CAP, has substantially lowered nitrogen and phosphorus surpluses and leaching of nitrogen from the root zone shown in the Box figures below.

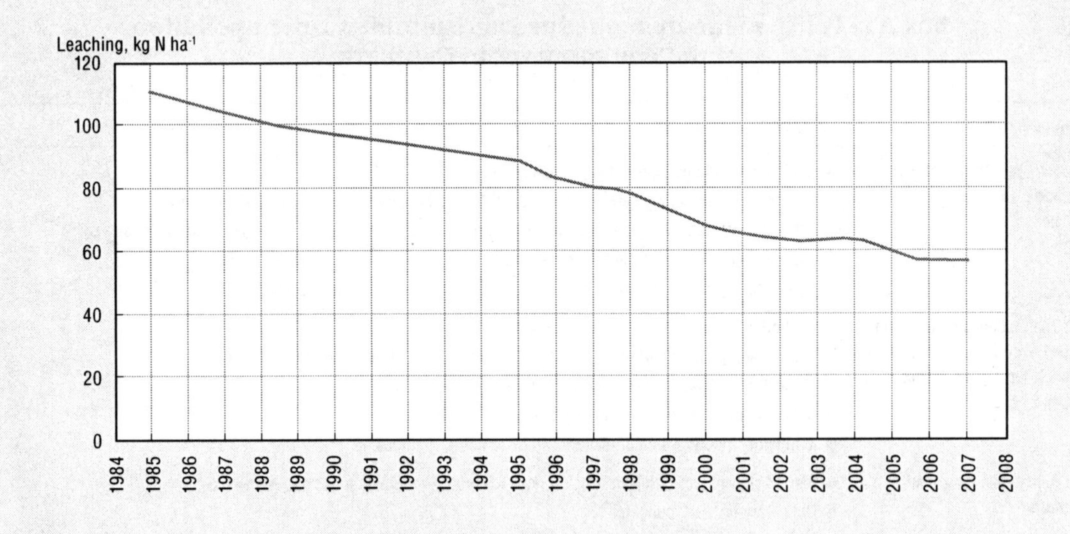

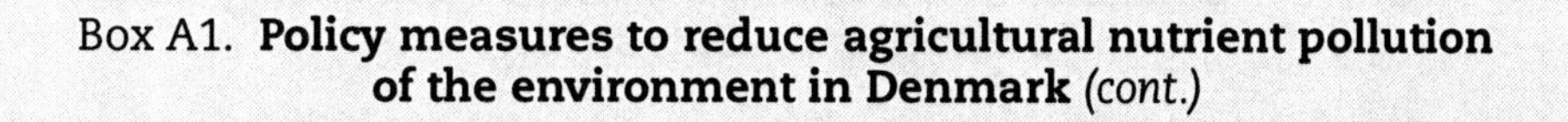

Box A1. **Policy measures to reduce agricultural nutrient pollution of the environment in Denmark** *(cont.)*

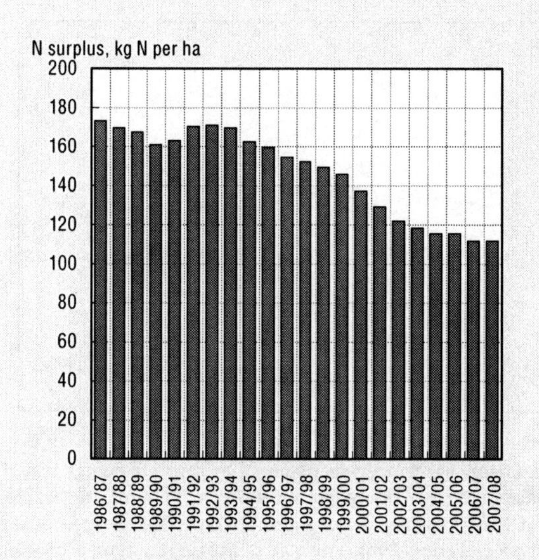

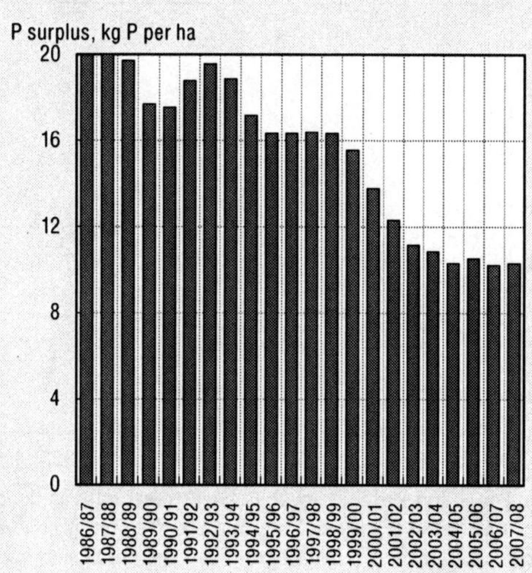

Source: Kronvang, B. et al. (2008), "Effects of Policy Measures Implemented in Denmark on Nitrogen Pollution of the Aquatic Environment", *Environmental Science and Policy*, Vol. 11, pp. 144-152; Maguire, R.O., G.H. Rubaek, B.E. Haggard and B.H. Foy (2009), "Critical Evaluation of the Implementation of Mitigation Options for Phosphorus from Field to Catchment Scales", *Journal of Environmental Quality*, Vol. 38, pp. 1989-1997; and Vinther, F.P. and C.D. Borgesen (2010), *Nutrient Surplus as a Tool for Evaluating Environmental Action Plans in Denmark*, presented at the OECD Workshop on agri-environmental indicators, March, Leysin, Switzerland, see OECD website: *www.oecd.org/agriculture/env/indicators/workshop*.

Switzerland: Improvement of input/output efficiency

Switzerland's indicators of the efficiency of the use of inputs, in particular nitrogen and energy, have not shown any improvement over time (Figure A.1). For this reason, the Federal Government decided that, from 2008, farmer support would be allocated with the purpose of improving the utilisation of natural resources in agriculture (regional programmes for the sustainable use of natural resources). The target areas are: the resources necessary for agricultural production, such as nitrogen, phosphorus and energy, the optimisation of plant protection; and increased protection and more sustainable use of soil, biodiversity in agriculture and the landscape.

European Union: Using indicators for evaluating and monitoring Rural Development Policy

Agri-environmental indicators are being used in the evaluation and monitoring of the **Rural Development Policy (RDP) of the European Union**. The objectives of the RDP for the period 2007-13, as defined in the Council Regulation No. 1698/2005, include three thematic axes:

1. improving the competitiveness of the agricultural and rural sector;
2. improving the environment and the countryside;
3. improving the quality of life in rural areas and encouraging diversification of the rural economy.

To reach these objectives, each country defines a national rural development programme which describes the measures to be undertaken. The evaluation process of the RDP includes a *mid-term* and an *ex-post* evaluation exercise. In order to ensure consistency

Figure A.1. **Efficiency of nitrogen, phosphorus and energy use, Switzerland, 1990-2006**

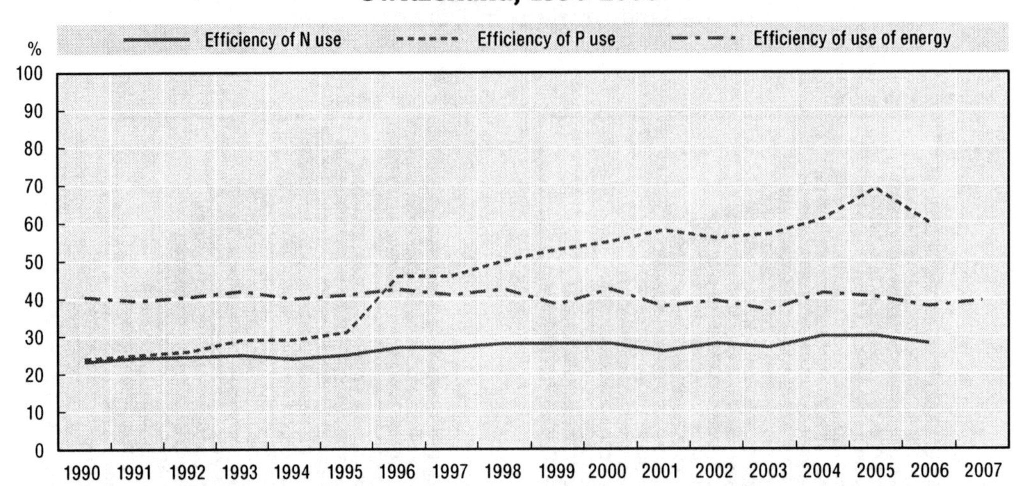

Notes: For N and P efficiency, the OSPAR method was used. Energy efficiency was obtained by dividing energy use of agriculture (fossil energy and uranium) by the energy contained in the agricultural products which are human digestible.
Source: Decrausaz, B. (2010), *Agri-Environmental Monitoring: A Tool for Evaluation and Support of Decision-Making for Swiss Agricultural Policy*, Office fédéral de l'Agriculture, Unité de direction Évaluation et stratégie, Secteur écologie, Switzerland, Paper presented to the OECD Workshop on agri-environmental indicators, March, Leysin, Switzerland, see OECD website: *www.oecd.org/agriculture/env/indicators/workshop*.

StatLink http://dx.doi.org/10.1787/888932793357

across countries, the evaluation process is undertaken under a common framework, the *Common Monitoring and Evaluation Framework* (CMEF). The CMEF proposes a list of environmental indicators that can be used as part of the evaluation exercise.

While the methodology and dataset used can vary in detail, most of these indicators closely correspond to the OECD set of agri-environmental indicators: population of farmland birds, gross nutrient balance, water withdrawals, water quality, etc. (see Table A.1 presenting a selection of the CMEF indicators). Two of the CMEF indicators, i.e. population of farmland birds and gross nutrient balance are *lead indicators*, i.e. indicators that countries should at least report in their evaluation. The CMEF guidance document also indicates that the gross nitrogen and phosphorous balances should be calculated using the OECD/Eurostat methodology (OECD, 2012a; and 2012b).

OECD work on Green Growth

The world economic crisis that began in 2008 has convinced many countries that a different kind of economic growth is needed. In response, many governments are putting in place measures aimed at a green recovery. Together with innovation, going green can be a long-term driver for economic growth, through, for example, investing in renewable energy and improved efficiency in the use of energy and materials.

By analysing economic and environmental policies together, by looking at ways to spur eco-innovation and by addressing other key issues related to a transition to a greener economy such as jobs and skills, investment, taxation, trade and development, the OECD is undertaking analysis to show the way to make a cleaner low-carbon economy compatible with green growth (OECD, 2011a). The OECD's agri-environmental indicators are contributing toward the OECD's work on Green Growth (OECD, 2011b).

Table A.1. **A selection of agri-environmental indicators used for the evaluation and monitoring of the Rural Development Policy in the European Union**

Indicator	Measurement
Land cover[1]	% area in agricultural/forest/natural/artificial classes
Biodiversity: Population of farmland birds[2]	Trends of index of population of farmland birds
Biodiversity: High nature value farmland and forestry[2]	Utilised Agriculture Area (UAA) of high nature value farmland
Biodiversity: Tree species composition	Distribution of species group by area of FOWL (% coniferous/% broadleaved/% mixed)
Water quality: Gross nutrient balance[2]	Surplus of nitrogen in kg/ha
	Surplus of phosphorous in kg/ha
Water pollution by nitrates and pesticides	Annual trends in the concentrations of nitrate and pesticides in ground and surface waters
Water quality[1]	% territory designated as nitrate vulnerable zone
Water use[1]	% irrigated UAA
Soil: Areas at risk of soil erosion	Areas at risk of soil erosion (t/ha/year)
Soil: Organic farming	UAA under organic farming
Climate change: Production of renewable energy from agriculture and forestry[2]	Production of renewable energy from agriculture (ktoe)
	Production of renewable energy from forestry (ktoe)
Climate change: UAA devoted to renewable energy	UAA devoted to energy and biomass crops
Climate change/air quality: Gas emissions	Emissions of greenhouse gases and ammonia from agriculture

1. These indicators are context related baseline indicators.
2. These indicators are lead indicators that a country should at least present.
Source: European Commission (2006), Directorate General for Agriculture and Rural Development.

Policies that promote green growth need to be based on a good understanding of the determinants of green growth and the related trade-offs or synergies. They also need to be supported with appropriate information to monitor progress and gauge results. In this context, the OECD report *Towards Green Growth: Monitoring Progress* (OECD, 2011c) proposes a set of multi-sector green growth indicators at the national level. These green growth indicators are embedded in a conceptual framework and selected according to well specified criteria, and accompanies the OECD Green Growth Strategy (OECD, 2011a).

In the OECD report on green growth monitoring, the OECD's nutrient and phosphorous balances indicators in agriculture are used to build partial environmental productivity indicators at the sector level for OECD countries (Figures 4.6 and 4.7). AEIs can thus play an important role as a basis for the development of a set of green growth indicators based on internationally comparable data. The broad green growth indicators are a starting point to be further elaborated by OECD as new data become available and concepts evolve.

The OECD report *Food, Agriculture and Green Growth* (OECD, 2011b) also presents some other examples of partial environmental and resource productivity indicators at the sector level using the OECD agri-environmental indicators. These indicators illustrate some specific issues by relating the evolution of agricultural production to the evolution of a particular agri-environmental indicator such as land area, agricultural water withdrawals, greenhouse gas emissions and nutrient balances. The report underlines that there are no existing indicators for the food and agriculture sector that, taken together, can track progress towards a global, comprehensive green growth indicator at this stage.

Economic, agri-environmental, natural resource stocks and social indicators exist, but are at various stages of development. In particular for agri-environmental and natural resource stocks, there are methodological, measurement and data availability problems. In the longer run, the development of resource intensity indicators are at different stages of development, and when possible the assessment of environmental externalities could help in assessing progress towards green growth in food and agriculture. When valuing positive and negative externalities generated by agriculture, agri-environmental indicators can provide a robust, recognised and useful basis for assisting policy decision making.

OECD Economic Surveys and Environmental Performance Reviews

Agri-environmental indicators are also used in two regular OECD publications: the OECD *Environmental Performance Reviews* and the OECD *Country Economic Surveys*. The utilisation of agri-environmental indicators in these reports, together with other sectoral datasets, facilitates policy monitoring and evaluation.

The OECD Environmental Performance Reviews "provide independent assessments of countries' progress in achieving their environmental policy objectives in order to help improve individual and collective environmental performance" (OECD website: *www.oecd.org/env/countryreviews*). Since their beginning in 1992, more than 60 Environmental Country Review exercises have been undertaken.

A recent review, the OECD ***Environmental Performance Review of Germany*** (OECD, 2012), makes use of several OECD agri-environmental indicators: nutrients surplus, agricultural greenhouse gas and ammonia emissions, and pesticide sales. Concerning nutrients surplus, the report notes that "the nitrogen surplus, at 100 kg per hectare, is still high [...]. The nitrate threshold (50 mg/l NO_3) was exceeded at 15% of monitoring sites" and that "several measures taken to improve the environmental performance of agriculture [...] helped reduce concentrations of phosphorus and nitrates in the main German rivers, although at a slower pace than in the 1990s." On pesticides, the report mentions that "the number of samples detecting pesticides above the threshold value decreased by nearly 50% between 1996-2000 and 2006-08" and recommends Germany to "pursue efforts to develop water quality monitoring, particularly for pesticides and nutrients in groundwater and lakes." The report illustrates how agri-environmental indicators can contribute to enrich environmental policy analysis and evaluation.

A summary of the use of agri-environmental indicators in recent OECD *Environmental Performance Reviews* and *Economic Country Surveys* is provided in Table A.2.

Table A.2. **Use of agri-environmental indicators in recent OECD Country *Environmental Performance Reviews* and *Economic Surveys***

	Type of report	Date of publication	Agri-environmental indicators used in the report
Germany	EPR	2012	GHG, ammonia, nutrient balance, water pollution (pesticides)
Slovenia	EPR	2011	Soil erosion, water use, water pollution, organic farming, GHG and ammonia
Israel	EPR	2011	Land use, water use, water pollution
New Zealand	ES	2011	GHG emissions, nutrient balance, pesticide use, direct on-farm energy consumption
France	ES	2011	GHG emissions, water quality
Spain	ES	2010	Irrigation water use, irrigated area, irrigation water application rate

EPR: Environmental Performance Review; ES: Economic Survey.

OECD *Country Economic Surveys* are published every two years for each OECD country, with the purpose to "identify the main economic challenges faced by the country and analyses policy options to meet them" (OECD, 2012, *www.oecd.org/eco/surveys*). In several cases, agri-environmental indicators have been used as part of the policy monitoring and evaluation in *Country Economic Surveys*, as illustrated by the recent examples of **France**, **New Zealand** and **Spain**, discussed below.

In the OECD ***Economic Surveys: France*** (OECD, 2011d), Chapter 4 examines *France's environmental policies: internalising global and local externalities*, in particular, water quality issues related to the use of fertilisers and pesticides in agriculture. OECD agri-environmental indicators reveal high level of pesticide use and nitrogenous fertiliser use by the agricultural sector.

The report notes that "the presence of pesticides was detected in 91% of river water and 59% of groundwater observation points. The pesticide content of water was higher than allowed by existing environmental standards in 11% and 18% of the respective observation points." Nitrogen pollution also represents a significant environmental issue for certain aquifers, with an excess of "the maximum admissible concentration of 50 mg/l (Groundwater Directive of 2006), above which water is considered undrinkable, in 6% of the observation sites in 2007 up from 4% in 1997." On the basis of this analysis, the report recommends to reinforce environmental policies targeted at these issues by developing either an input tax or a quota system on fertiliser and pesticide use.

The OECD (2011e) ***Economic Survey: New Zealand*** 2011, made an assessment of green growth and climate change in New Zealand. The report, drawing on the OECD's AEIs included a review of water quality, noting that nitrogenous effluent from agricultural fertiliser and animal urine seeps through the soil and into surrounding lakes and rivers where it nourishes the growth of algae, which in turn diminishes the quality and aesthetic value of lakes, while harbouring waterborne diseases. Biodiversity is harmed, as the same nitrogen leaching causes eutrophication of waterways.

The report notes that the impairment of water flow in rivers and of aquifer levels during droughts and increased abstractions from irrigation systems has exacerbated such quality problems insofar as the absorptive capacity of the water decreases. Recreational water uses that are fundamental to the tourism industry and New Zealand's lifestyle alike, increasingly collide with agricultural and community uses. Even so, New Zealand's agricultural nitrogen balance, while increasing, is still much lower than some other OECD countries on account of the extensive pastoralism practiced and the absence of production and input support. The report concluded that New Zealand should continue to develop measurement of water quality via evolving national guidelines and apply pollution-rights trading to address water pollution.

In the OECD ***Economic Survey: Spain*** (OECD, 2010), a substantial part of the analysis is devoted to the issue of sustainable water management, in particular groundwater drawing on the OECD AEIs (see Chapter 4 of the report, *Policies towards a sustainable use of water*). The report underlines the importance of agriculture in the share of total water withdrawals, and the tendency over the last decade of increasing groundwater withdrawals. In addition, nitrate pollution is also a significant threat to water quality, in particular for groundwater.

Assessing the global biodiversity outlook

The Secretariat of the Convention on Biological Diversity (2010), *Global Biodiversity Outlook* 3, in its assessment of global biodiversity trends and outlook, examined the pressure on biodiversity from water pollution, drawing on the OECD nitrogen balance indicator. "Pollution from nutrients (nitrogen and phosphorous) and other sources is a continuing and growing threat to biodiversity in terrestrial, inland water and coastal ecosystems. In inland water and coastal ecosystems, the build-up of phosphorous and nitrogen, mainly through run-off from cropland and sewage pollution, stimulates the growth of algae and some forms of bacteria, threatening valuable ecosystem services in systems such as lakes and coral reefs, and affecting water quality. It also creates 'dead zones' in oceans, generally where major rivers reach the sea. In these zones, decomposing algae use up oxygen in the water and leave large areas virtually devoid of marine life."

"The number of reported dead zones has been roughly doubling every ten years since the 1960s, and by 2007 had reached around 500. While the increase in nutrient load is among the most significant changes humans are making to ecosystems, policies in some regions are showing that this pressure can be controlled and, in time, reversed. Among the most comprehensive measures to combat nutrient pollution is the European Union's *Nitrates Directive*. The average nitrogen balance per hectare of agricultural land (the amount of nitrogen added to land as fertiliser (and livestock manure), compared with the amount used up by crops and pasture) for selected European countries is shown in (Figure A.2). The reduction over time in some countries implies improved efficiency in the use of fertiliser, and therefore a reduced risk of damage to biodiversity through nutrient run-off".

Figure A.2. **Nitrogen surplus balance, OECD European countries, 1990-2004**

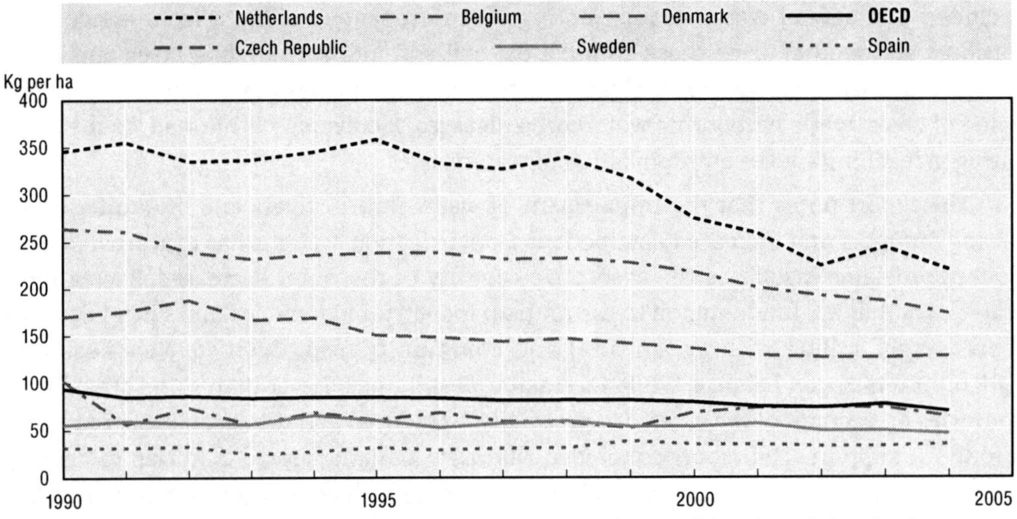

Source: The Secretariat of the Convention on Biological Diversity (2010), *Global Biodiversity Outlook 3*, drawing on OECD (2008), *Environmental Performance of Agriculture Since 1990*.

StatLink http://dx.doi.org/10.1787/888932793376

Research community use of agri-environmental indicators

The wider research community is also making use of the OECD's AEIs and related publications, as summarised for selected papers and reports in this annex. The use of AEIs in the research community mainly concerns the following activities (and a few illustrative examples are provided in the discussion below):

- developing agri-environmental indicators drawing on the OECD agri-environmental indicators;
- utilising OECD agri-environmental indicators for the analysis and evaluation of the environmental performance of agriculture; and
- drawing on the OECD AEI structure to undertake policy analysis.

Bouwman et al. (2009) developed spatially explicit global trends in nitrogen and phosphorus soil balance over time by using IMAGE (Integrated Model to Assess the Global Environment). To verify their model, they also compared their nitrogen and phosphorus balances for the year 2000 with country estimates from OECD (2008) for 29 OECD countries, which revealed a strong correlation with the OECD nutrient balances. By using different scenarios in their model, they predicted the future global balances of nitrogen and phosphorus balances.

Calculations were undertaken by Potter et al. (2010) of spatially explicit fertiliser inputs of nitrogen and phosphorus by combining various national data, and also manure inputs of nitrogen and phosphorus by using international data including the OECD AEIs. They identified that a few hot spots contribute to the accumulation of nutrients and water quality problems.

In a study by Hoang and Alauddin (2010) investigated nitrogen flows and balance of OECD countries from 1985 to 2003. They chose three indicators to assess relative environmental performance: the eco-environmental indicator; system nutrient efficiency; and the nutrient balance normalised by agricultural land. Although the OECD (2008) used the nutrient balance normalised by agricultural land, Hoang and Alauddin used all three indicators for comparing the performance among OECD countries.

Selective examples of research literature (since 2008) which draw on the OECD's agri-environmental indicators

Abrantes, N., R. Pereira and F. Gonçalves (2010), "Occurrence of Pesticides in Water, Sediments, and Fish Tissues in a Lake Surrounded by Agricultural Lands: Concerning Risks to Humans and Ecological Receptors", *Water, Air, & Soil Pollution*, Vol. 212, No. 1-4, pp. 77-88.

Adl, S., D. Iron and T. Kolokolnikov (2011), "A Threshold Area Ratio of Organic to Conventional Agriculture Causes Recurrent Pathogen Outbreaks in Organic Agriculture", *Science of the Total Environment*, Vol. 409, No. 11, pp. 2192-2197.

Balat, H. and C. Öz (2008). "Challenges and Opportunities for Bio-Diesel Production in Turkey", *Energy, Exploration & Exploitation*, Vol. 26, No. 5, pp. 327-346.

Bassanino, M., D. Sacco, L. Zavattaro and C. Grignani (2011), "Nutrient Balance as a Sustainability Indicator of Different Agro-environments in Italy", *Ecological Indicators*, Vol. 11, No. 2, pp. 715-723.

Carew, R. (2010), "Ammonia Emissions from Livestock Industries in Canada: Feasibility of Abatement Strategies", *Environmental Pollution*, Vol. 158, No. 8, pp. 2618-2626.

Darnhofera, I., J. Fairweatherb and H. Mollerc (2010), "Assessing a Farm's Sustainability: Insights from Resilience Thinking", *International Journal of Agricultural Sustainability*, Vol. 8, No. 3, pp. 186-198.

Gasparatos, A. (2010), "Resource Consumption in Japanese Agriculture and its Link to Food Security", *Energy Policy*, Vol. 39, pp. 1101-1112.

Gallego-Ayala, J. and J.A. Gómez-Limón (2009), "Analysis of Policy Instruments for Control of Nitrate Pollution in Irrigated Agriculture in Castilla y León, Spain", *Spanish Journal of Agricultural Research*, Vol. 7, No. 1, pp. 24-40.

Grizzetti, B., F. Bouraoui and A. Aloe (2012), "Changes of Nitrogen and Phosphorus Loads to European Seas", *Global Change Biology*, Vol. 18, No 2, pp. 769-782.

Hadjikakoua, M., P.G. Whiteheada, L. Jina, M. Futterc, P. Hadjinicolaoud and M. Shahgedanovae (2011), "Modelling Nitrogen in the Yesilirmak River Catchment in Northern Turkey: Impacts of Future Climate and Environmental Change and Implications for Nutrient Management", *Science of the Total Environment*, Vol. 409, No. 12, pp. 2404-2418.

Hoang, V.N. and D.S.P. Rao (2010), "Measuring and Decomposing Sustainable Efficiency in Agricultural Production: A Cumulative Energy Balance Approach", *Ecological Economics*, Vol. 69, No. 9, pp. 1765-1776.

Hoang, V.N. (2011), "Measuring and Decomposing Changes in Agricultural Productivity, Nitrogen Use Efficiency and Cumulative Energy Efficiency: Application to OECD Agriculture", *Ecological Modelling*, Vol. 222, No. 1, pp. 164-175.

Hoang, V.N. and M. Alauddin (2011), "Analysis of Agricultural Sustainability: A Review of Energy Methodologies and their Application in OECD Countries", *International Journal of Energy Research*, Vol. 35, No. 6, pp. 459-476.

Hoang, V.N. and M. Alauddin (2011), "Input-Orientated Data Envelopment Analysis Framework for Measuring and Decomposing Economic, Environmental and Ecological Efficiency: An Application to OECD Agriculture", *Environmental and Resource Economics*.

Kaygusuz, K. (2010), "Sustainable Energy, Environmental and Agricultural Policies in Turkey", *Energy Conversion and Management*, Vol. 51, pp. 1075-84.

Mishima, S., A. Endo and K. Kohyama (2010), "Recent Trends in Phosphate Balance Nationally and by Region in Japan", *Nutrient Cycling in Agroecosystems*, Vol. 86, No. 1, pp. 69-77.

Mishima, S., A. Endo and K. Kohyama (2010), "Nitrogen and Phosphate Balance on Crop Production in Japan on National and Prefectural Scales", *Nutrient Cycling in Agroecosystems*, Vol. 87, No. 2, pp. 159-173.

Muñoz, I., L. Milà-i-Canals and A.R. Fernández-Alba (2010), "Life Cycle Assessment of Water Supply Plans in Mediterranean Spain", *Journal of Industrial Ecology*, Vol. 14, No. 6, pp. 902-918.

Novotny, V., X. Wang, A.J. Englande, D. Bedoya, L. Promakasikorn and R. Tirado (2010), "Comparative Assessment of Pollution by the Use of Industrial Agricultural Fertilizers in Four Rapidly Developing Asian Countries", *Environment, Development and Sustainability*, Vol. 12, No. 4, pp. 491-509.

Parris, K (2011), "Impact of Agriculture on Water Pollution in OECD Countries: Recent Trends and Future Prospects", *International Journal of Water Resources Development*, Vol. 27, No. 1, pp. 33-52.

Pechera, C., E. Tassera and U. Tappeinera (2011), "Definition of the Potential Treeline in the European Alps and its Benefit for Sustainability Monitoring". *Ecological Indicators*, Vol. 11, No. 2, pp. 438-447.

Prabodanie, R.A.R., J.F. Raffensperger, E.G. Read and M.W. Milke (2011), "LP Models for Pricing Diffuse Nitrate Discharge Permits", *Annals of Operations Research*, forthcoming.

Rivetta, M.O., S.R. Bussb, P. Morganb, J.W.N. Smith and C.D. Bemment (2008), "Nitrate Attenuation in Groundwater: A Review of Biogeochemical Controlling Processes", *Water Research*, Vol. 42, No. 16, pp. 4215-4232.

Rüdissera, J., E. Tasser and U. Tappeinera (2012), "Distance to Nature – A New Biodiversity Relevant Environmental Indicator Set at the Landscape Level", *Ecological Indicators*, Vol. 15, No. 1, pp. 208-216.

Sheppard, S.C., S. Bittman, M.L. Swift, M. Beaulieu and M. Sheppard (2011), "Ecoregion and Farm Size Differences in Dairy Feed and Manure Nitrogen Management: A Survey", *Canadian Journal of Animal Science*, Vol. 91, No. 3, pp. 459-473.

Sieber, S., D. Pannell, K. Muller, K. Holm-Muller, P. Kreinsd and V. Gutschee (2010), "Modelling Pesticide Risk: A Marginal Cost-benefit Analysis of an Environmental Buffer-zone Programme", *Land Use Policy*, Vol. 27, No. 2, pp. 653-661.

Spiess, E. (2011), "Nitrogen, Phosphorus and Potassium Balances and Cycles of Swiss Agriculture From 1975 to 2008", *Nutrient Cycling in Agroecosystems*, Vol. 91, No. 3, pp. 351-365.

Sugito, T., K. Yoshida, M. Takebe, T. Shinano and K. Toyota (2010), "Soil Microbial Biomass Phosphorus as an Indicator of Phosphorus Availability in a Gleyic Andosol", *Soil Science and Plant Nutrition*, Vol. 56, No. 3, pp. 390-398.

Swink, S.N., Q.M. Ketterings, L.E. Chase, K.J. Czymmek and M.E. van Amburgh (2011), "Nitrogen Balances for New York State: Implications for Manure and Fertilizer Management", *Journal of Soil and Water Conservation*, Vol. 66, No. 1, pp. 1-17.

Van Hoi, P., A.P.J. Mol and P.J.M. Oosterveer (2009), "Market Governance for Safe Food in Developing Countries: The Case of Low-pesticide Vegetables in Vietnam", *Journal of Environmental Management*, Vol. 91, No. 2, pp. 380-388.

VanderZaag, A.C., S. Jayasundara and C. Wagner-Riddle (2011), "Strategies to Mitigate Nitrous Oxide Emissions from Land Applied Manure", *Animal Feed Science and Technology*, Vol. 166-167, pp. 464-479.

Vidal, T., N. Abrantes, A.M.M. Gonçalves and F. Gonçalves (2011), "Acute and Chronic Toxicity of Betanal® Expert and its Active Ingredients on Nontarget Aquatic Organisms from Different Trophic Levels", *Environmental Toxicology*.

Weaver, D.M. and M.T.F. Wong (2011), "Scope to Improve Phosphorus (P) Management and Balance Efficiency of Crop and Pasture Soils with Contrasting P Status and Buffering Indices", *Plant and Soil*, Vol. 349, No. 1-2, pp. 37-54.

de Wit, M., M. Londo and A. Faaij (2011), "Productivity Developments in European Agriculture: Relations to and Opportunities for Biomass Production", *Renewable and Sustainable Energy Reviews*, Vol. 15, No. 5, pp. 2397-2412.

References

Bouwman, A.F., A.H.W. Beusen and G. Billen (2009), "Human Alteration of the Global Nitrogen and Phosphorus Soil Balances for the Period 1970-2050", *Global Biogeochemical Cycles*, Vol. 23.

Convention on Biological Diversity (2010), *Global Biodiversity Outlook 3*, CBD, Montreal, Canada.

Hoang, V.N. and M. Alauddin (2010), "Assessing the Eco-environmental Performance of Agricultural Production in OECD Countries: The Use of Nitrogen Flows and Balance", *Nutrient Cycling in Agroecosystems*, Vol. 87, No. 3, pp. 353-368.

OECD (2012), *OECD Environmental Performance Reviews: Germany 2012*, OECD Publishing.

OECD (2011a), *Towards Green Growth*, OECD Publishing, *www.oecd.org/greengrowth*.

OECD (2011b), *Food, Agriculture and Green Growth*, OECD Publishing, *www.oecd.org/greengrowth*.

OECD (2011c), *Towards Green Growth: Monitoring Progress OECD Indicators*, OECD Publishing, *www.oecd.org/greengrowth*.

OECD (2011d), *OECD Economic Surveys: France 2011*, OECD Publishing.

OECD (2011e), *OECD Economic Surveys: New Zealand 2011*, OECD Publishing.

OECD (2010), *OECD Economic Surveys: Spain 2010*, OECD Publishing.

OECD (2008), *Environmental Performance of Agriculture at a Glance*, OECD Publishing, *www.oecd.org/tad/sustainable-agriculture/agri-environmentalindicators.htm*.

OECD/EUROSTAT (2012a), *OECD/EUROSTAT Nitrogen Balance Handbook*, *www.oecd.org/tad/sustainable-agriculture/agri-environmentalindicators.htm*.

OECD/EUROSTAT (2012b), *OECD/EUROSTAT Phosphorus Balance Handbook*, *www.oecd.org/tad/sustainable-agriculture/agri-environmentalindicators.htm*.

Potter, P., N. Ramankutty, E. Bennett and S. Donner (2010), "Characterizing the Spatial Patterns of Global Fertilizer Application and Manure Production", *Earth Interactions*, Vol. 14, No. 2, pp. 1-22.

ORGANISATION FOR ECONOMIC CO-OPERATION AND DEVELOPMENT

The OECD is a unique forum where governments work together to address the economic, social and environmental challenges of globalisation. The OECD is also at the forefront of efforts to understand and to help governments respond to new developments and concerns, such as corporate governance, the information economy and the challenges of an ageing population. The Organisation provides a setting where governments can compare policy experiences, seek answers to common problems, identify good practice and work to co-ordinate domestic and international policies.

The OECD member countries are: Australia, Austria, Belgium, Canada, Chile, the Czech Republic, Denmark, Estonia, Finland, France, Germany, Greece, Hungary, Iceland, Ireland, Israel, Italy, Japan, Korea, Luxembourg, Mexico, the Netherlands, New Zealand, Norway, Poland, Portugal, the Slovak Republic, Slovenia, Spain, Sweden, Switzerland, Turkey, the United Kingdom and the United States. The European Union takes part in the work of the OECD.

OECD Publishing disseminates widely the results of the Organisation's statistics gathering and research on economic, social and environmental issues, as well as the conventions, guidelines and standards agreed by its members.

OECD PUBLISHING, 2, rue André-Pascal, 75775 PARIS CEDEX 16
(51 2013 01 1 P) ISBN 978-92-64-18115-1 – No. 60589 2013-02

Lightning Source UK Ltd.
Milton Keynes UK
UKOW02f1112210813

215736UK00002B/6/P

9 789264 181151